A LATE FORMATIVE IRRIGATION SETTLEMENT
BELOW MONTE ALBÁN

A Late Formative Irrigation Settlement below Monte Albán

Survey and Excavation on the Xoxocotlán Piedmont, Oaxaca, Mexico

by Michael J. O'Brien, Roger D. Mason,
Dennis E. Lewarch, and James A. Neely

INSTITUTE OF LATIN AMERICAN STUDIES
THE UNIVERSITY OF TEXAS AT AUSTIN

INSTITUTE OF LATIN AMERICAN STUDIES
William P. Glade, *Director*
Robert M. Malina, *Associate Director*

International Standard Book Number 0-292-74628-8
Library of Congress Card Catalogue Number 80-85398
© 1982 by the Institute of Latin American Studies,
The University of Texas at Austin. All rights reserved.

PRINTED BY THE UNIVERSITY PRINTING DIVISION OF
THE UNIVERSITY OF TEXAS AT AUSTIN
MANUFACTURED IN THE UNITED STATES OF AMERICA

Distributed for the Institute of Latin
American Studies by:
    University of Texas Press
    P. O. Box 7819
    Austin, Texas 78712

Contents

Tables

Figures

Plates

Preface

Although our appreciation is extended to all involved with our
project, we would like to single out a few of these people and institutions for special recognition. Our gratitude goes to Kent V. Flannery
of the University of Michigan, founder and director of the Valley of
Oaxaca Human Ecology Project, who loaned us much of our equipment, and
to Richard E. Blanton of Purdue University, director of the Valley of
Oaxaca Settlement Pattern Project, under whose permit we worked. We
would also like to thank Frank Hole of Rice University for loaning
us other items of equipment.

The Institute of Latin American Studies at the University of Texas
at Austin provided funding by means of a Summer Faculty Research Grant
to Neely and an E. D. Farmer Fellowship to O'Brien. Without this
financial assistance we would have been unable to continue work in 1973.
Lewarch acknowledges support provided by an NSF Traineeship at the
University of Washington, which allowed him to carry out computer
analysis of the data. Mason was an NDEA Title IV Fellow during the
period of fieldwork.

The following people volunteered their time and effort in the field:
Ira Beckerman, Margie Lohse, Cyndi Oliver, Janet Mitchell, Dirk Hood,
Bruce Byland, Annabelle Hoffmeister and Mollie Conrades. In addition,
Chris Atkins, Barbara Burger, and Mollie Conrades drew many of the
artifacts. Richard Wilkerson of the State University of New York
at Albany graciously analyzed the skeletal material. Jerry V. Jermann
of the University of Washington provided immeasurable assistance in
using the SYMAP computer mapping system.

Residents of Oaxaca to whom we owe gratitude include Cecil Welte,
of the Oficina de Estudios de Humanidad del Valle de Oaxaca and Manuel
Esparza, regional director of the Instituto Nacional de Antropología
e Historia, who helped us cut through the "red tape" involved in working
in the Valley. Our largest debt of gratitude goes to Marcus C. Winter,

a founding member of the Oaxaca Project. He was a source both of
information and of inspiration to us. Not only did he help us untangle
legal matters, but he also showed us the changes he proposed for the
local ceramic sequence. This took a great deal of his time, and
most surely tried his patience upon occasion, yet he unflaggingly and
cheerfully continued to assist us. We would not have been able to
interpret our results to the degree that we have, had not Marc lent
his assistance.

Finally, there are two people whom we find difficulty in thanking--
for it is like thanking ourselves. John and Melissa Keane were codirec-
tors of the project and were involved with it from July 1972 until
August 1973. They share credit with us in every way for the accom-
plishments of this project. Their knowledge of archaeology and their
friendliness in and out of the field helped the project run smoothly.
For personal reasons they decided not to coauthor this report with us,
but many of the ideas incorporated herein can be attributed in part
to them.

A LATE FORMATIVE IRRIGATION SETTLEMENT
BELOW MONTE ALBÁN

1. Introduction

This monograph is a report of results and interpretations of archaeological fieldwork undertaken in the summer of 1973 in the vicinity of a Prehispanic irrigation system on the piedmont below and to the east of the site of Monte Albán in Oaxaca, Mexico. The irrigation system was originally located by James A. Neely (1972), who carried out a reconnaissance of the Monte Albán area in 1971 designed to locate water-control features. This reconnaissance was undertaken in association with the Monte Albán Survey, directed by Richard E. Blanton (1978).

The irrigation system located by Neely consisted of a mortared masonry dam which impounded runoff from Monte Albán in a barranca on its eastern slope, and a canal about two kilometers long which ran from the dam onto a low ridge and into the piedmont zone near the town of Xoxocotlán. The upper piedmont zone around the canal was terraced, presumably for agricultural purposes in conjunction with the canal. Examination of sherds from the canal fill indicated a Late Formative date for the system (Neely and O'Brien 1973). The presence of several small mounds and a distribution of sherds in the area indicated that there was an associated settlement. The system is further described in chapter V.

The discovery of a Late Formative irrigation community associated with Monte Albán, which became a Classic period center of major proportions, suggested the study of two major questions during the 1973 field work (summarized in Mason et al. 1977). The first concerned the effect of an irrigation system on internal community patterning with respect to the canal. That is, did the presence of a canal distort the usual Oaxacan hamlet or village pattern of a contiguous, roughly circular area of residences, as described by Marcus Winter (1976) and Robert Drennan (1976)? If so, one might hypothesize a more linear

4

arrangement, with settlement strung out along the canal and higher-
status residences nearer the water source for purposes of control.

The second question concerns the relationship of the irrigation
system and its associated settlement to Monte Albán proper. Although
thoughts of Wittfogel and the hydraulic hypothesis (Wittfogel 1957)
immediately came to mind, it was obvious that the Xoxocotlán irrigation
system was not of sufficient magnitude or scale (50 hectares of irrigable area) to have been a "prime-mover" in the development of Monte
Albán, which according to Blanton (1978), became the administrative
center of a "military confederacy" by 200 B.C. The hydraulic hypothesis
is a restatement of Wittfogel's model of hydraulic society in evolution-
ary terms. The hypothesis states, in brief, that the managerial
requirements of large-scale irrigation agriculture lead to development
of more complex forms of political organization, especially the state
(see Sanders 1968 for an application of the hydraulic hypothesis to
Teotihuacán).

Even though it was not anticipated that study of the Xoxocotlán
irrigation system would provide data directly contributing to a study
of the causes of the growth of Monte Albán, it was still of interest
to determine what contribution the system could have made to the food
supply of Monte Albán. To do this, it was necessary to determine the
irrigable area of the system and estimate yields of the system in terms
of number of people it would support. This entailed study of carrying
capacity, using yields and consumption data from a piedmont irrigation
community in the present-day Valley of Oaxaca, with allowances made
for smaller corncob sizes in the past. This information is found in
chapter XIII.

It was also necessary to determine the extent of the residential
area in direct association with the canal system, to provide the basis
for an estimate of number of local consumers so that the amount of
surplus available for Monte Albán could be estimated. The determination
of residential area required field techniques which would produce
quantifiable and comparable data from all surface-collection units.
The specific field techniques adopted for surface collection in 1973
are described in chapter II. Basically, they consisted of using modern
field boundaries to define collection units and collecting all rim
sherds within each unit. The specific techniques employed had not,
to the authors' knowledge, been previously employed in Mesoamerica
(although they are generally similar to the techniques employed at
Teotihuacán). They constitute a level of surface-collection intensity

intermediate between that of regional survey and that employed when
collecting in small grid units within individual sites. This level of
intensity can be termed a zonal approach, in contrast to regional and
single-site approaches (Mason 1979). Therefore, work reported here
can also be seen as a test of field techniques. The results produced
by the surface-collection procedure employed are presented in chapter
XII, which consists of discussion of distributional patterns of sherds
collected for each phase.

As part of the test of the collection procedure, it was necessary to control for possible biases. These biases, or factors which
could have affected comparability of the surface collections, are
discussed in chapter XI. Some factors discussed include differences
in thoroughness of collection among crew members, level of interest
and degree of experience among crew members, and differences due to
size of collection area and small sample sizes, as well as problems
caused by vegetation. These possible biases are discussed in terms
of their effects on patterning of archaeological material. Once
effects of these possible biases have been assessed, it can be stated
whether the patterns observed in the collected material are due to
actual variation in distribution of archaeological classes or to
systematic bias in collection procedure. A discussion of biases is
usually neglected in archaeological reports, but it is felt by the
authors that factors affecting comparability of units should be considered and discussed, especially when quantitative comparisons
among units are sought. It is hoped that inclusion of this chapter
will stimulate others to explore the subject further.

Test excavations were also planned, to corroborate data from the
surface collections. Due to permit limitations, however, excavation
had to be limited to only two weeks and had to be undertaken before
surface collections were completed. Other areas would have been
chosen for excavation had it been possible to excavate after completion of the survey. The excavations did show, however, that relatively
high-density surface distributions did overlie subsurface features and
that the mound closest to the canal was built at the same time the
canal was in use. The excavations are described in chapter VI.

Ceramics and other artifacts recovered are described in chapters
VII and VIII. Ceramics were analyzed using modifications of the Caso,
Bernal, and Acosta (1967) system suggested by Marcus Winter. These
modifications consisted mainly of heavier emphasis on vessel form and
rim form as temporal indicators. These vessel forms and their

correspondence to the Caso, Bernal, and Acosta types are presented
in chapter VII. The differential distributions of vessel forms between an adjacent house and midden area are also discussed in that
chapter. However, since Winter's vessel-form classification system
was preliminary and based on personal communication only, no new formal
types are presented, and surface-collection tabulations by type could
not be presented here. Publication of these formal types by Winter
will be forthcoming.

It is realized that research presented here is only meaningful
when seen in the context of other, larger bodies of data from Oaxaca.
The recent publication of the Monte Albán Survey data by Blanton
(1978) and the various publications of Kent Flannery's group (Kirkby
1973, Lees 1973, Pires-Fereira 1975, Drennan 1976, Flannery, ed., 1970,
1976) help provide that context. The relationship of the small body
of data presented here to the results of these other projects is
discussed in the Conclusions. The Xoxocotlán Project was an outgrowth
of these other, larger projects, which are briefly described below.

RELATED ARCHAEOLOGICAL PROJECTS IN OAXACA

The Valley of Oaxaca is the largest valley in the Southern Highlands of Mesoamerica (figs. 1 and 2). At its center lies the site
of Monte Albán, a large Late Formative and Classic period center
which may have been the administrative center of a regional confederacy as early as 200 B.C. (Blanton 1978). Early research in Oaxaca
was concentrated at Monte Albán where Alfonso Caso directed excavations
which were designed primarily to recover complete artifacts from tombs.
The results of Caso's work, therefore, emphasized ceremonial and elite
segments of Monte Albán culture (Caso and Bernal 1952; Caso, Bernal,
and Acosta 1967).

The Valley of Oaxaca Human Ecology Project was initiated by
Flannery in 1966 and concentrated on exposure of Early and Middle
Formative period village activity areas through large horizontal
excavations. Through reconstruction of settlement-subsistence activities for various time periods, the general goal of constructing
"a model of Early Formative Society" (Flannery 1976a:8) could be
realized. While Flannery concentrated on San José Mogoté, a large
Early-Middle Formative regional center in the Etla arm of the Valley,
Winter excavated at Tierras Largas near Monte Albán (Winter 1970,
1972), and Robert Drennan (1976) explored Fábrica San José, a salt-

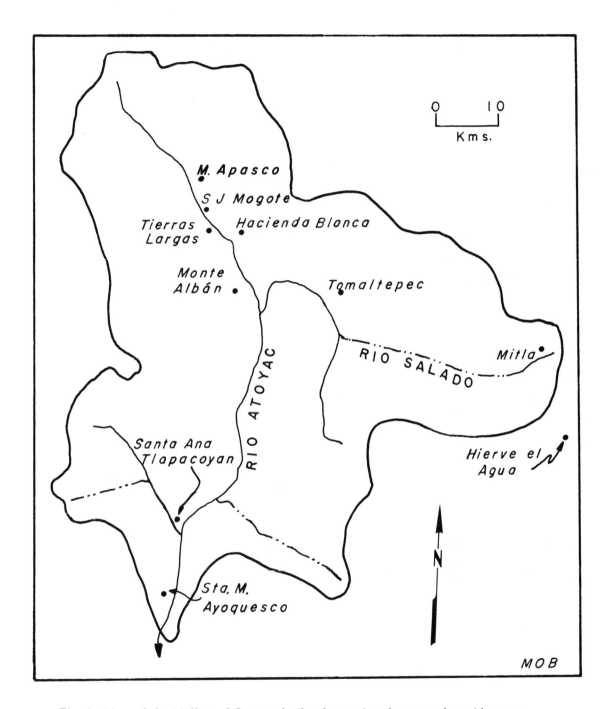

Fig. 1. Map of the Valley of Oaxaca indicating major sites mentioned in text
(adapted from Kirkby 1973).

8

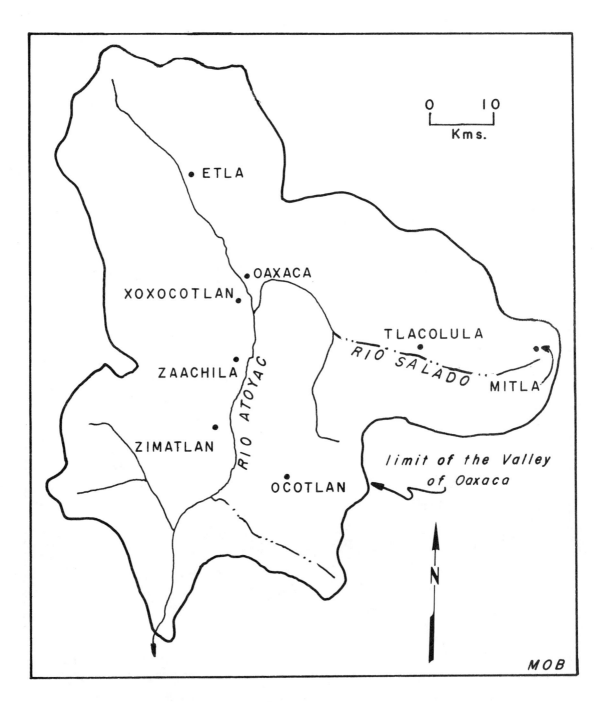

Fig. 2. Map of the Valley of Oaxaca showing main towns and other localities
mentioned in text.

producing site near San José Mogoté in the piedmont zone. Excavations were also undertaken at Huitzo in the Etla arm and at Abasolo and Tomaltepec in the Tlacolula arm. No excavations were carried out in the southern (Zaachila) arm. The work of Flannery and his associates provides background for the subsequent development of Monte Albán in the Late Formative.

In 1971, Blanton initiated the Valley of Oaxaca Settlement Pattern Project, a regional survey designed to locate, date, and determine the extent of all sites in the Valley (Blanton 1973). Techniques were similar to those employed by Jeffrey R. Parsons and Blanton in the Valley of Mexico regional surveys (Parsons 1971a, 1971b; Blanton 1972). The first phase of the project was the survey of the Etla arm of the Valley carried out by Dudley Varner (1974), followed by Stephen Kowalewski's (1976) survey of the central part of the Valley (the area surrounding Monte Albán). Blanton and Kowalewski have recently completed survey of the southern arm of the Valley (Kowalewski 1978).

As part of the Valley of Oaxaca Settlement Pattern Project, Blanton undertook the survey of Monte Albán, which was largely completed during the summers of 1971 and 1972. Approximately 2000 residential terraces were located and mapped on the slopes of the hills of Monte Albán, Monte Albán Chico, El Gallo, and Atzompa (fig. 3). Sherds were collected to date these features, in order to provide a picture of the occupation and growth of the site through time (Blanton 1978). One of the residential terraces was excavated by Winter (1974a, b) and is the only study of lower-status residences available for Monte Albán.

The Xoxocotlán Project developed directly from the Monte Albán Survey. Neely had conducted reconnaissance of the Monte Albán area which located the irrigation system on the piedmont east of Monte Albán, and O'Brien and Mason had both participated in the Monte Albán Survey. Thus, in the summer of 1973 it was decided to extend survey of the Monte Albán area onto the piedmont around the irrigation system, but employing more intensive collection techniques than those of the Monte Albán Survey (these are discussed further in chapter II).

The only other prehistoric irrigation system known in the Oaxaca area is that preserved in travertine in the hills above Mitla at Hierve el Agua (Neely 1967, 1970). However, this system is located outside the Valley of Oaxaca and is not associated with a major site. Thus, it contributes little to an understanding of developments in the

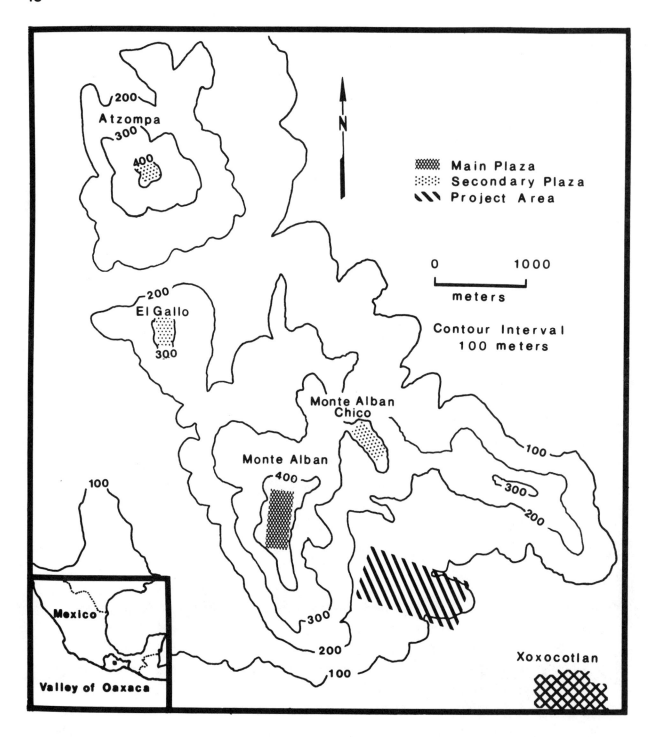

Fig. 3. Location of Xoxocotlán Survey Project area (adapted from Blanton 1973)

Valley.

The only previous work specific to the Xoxocotlán area was done by Marshall Saville in 1898. Saville (1899) excavated a mound complex on the east side of the Xoxocotlán-Cuilapan road, locating elaborate offerings. Based on the urns Saville illustrates, the mounds probably date to the Monte Albán III-B phase. Bernal (1965:793) states that Acosta at one time explored the Xoxocotlán area, but gives no further information. The Xoxocotlán area was also incorporated in Kowalewski's (1976) survey of the Central Valley.

2. Field Techniques

In order to assess utility of different field techniques, one must evaluate possibilities relative to the problem being considered. The first question to be asked about any technique is, will it help solve the research problem? After excluding those field operations which do not appear to be useful, one chooses among remaining alternatives on the basis of pragmatic considerations such as time, money, and personnel.

STRATEGIES

A number of possible surface-collection techniques of varying intensities were potentially useful for the problem at hand, i.e., quantitatively defining patterns in occupational densities pertaining to phases in the area around the irrigation canal. Surveys and their corresponding surface-collection techniques can be divided into three levels of intensity: regional, zonal, and site-intensive (Mason 1979).

Regional surveys cover the most area and employ the least-intensive collection techniques. The primary goal of regional surveys is location of sites in a region. Once located, site size and occupational density during various time periods are usually determined by making subjective, visual estimates of areal extent and density of sherds known to be diagnostic of various time periods. Surface collection in this type of survey is restricted to small grab samples made as a check on visual estimates.

Regional surveys in highland central Mexico have been widespread, so that much of central Mexico has now been surveyed for locations of archaeological sites. The pioneer of regional survey in central Mexico was William T. Sanders, who began the Teotihuacán Valley Survey in the early 1960s (Sanders 1965, 1970, 1975). Parsons, Blanton, and Sanders continued the regional surveys of the Basin of Mexico, and most

of it has now been completed (Parsons 1971, 1974; Blanton 1972).
Blanton moved to the Valley of Oaxaca and, in collaboration with
Varner and Kowalewski, has surveyed much of the Valley (Blanton 1973;
Varner 1974; Kowalewski 1976, 1978). Ronald Spores and Bruce E.
Byland have surveyed parts of the Mixteca Alta (Spores 1972; Byland
1978), and Kenneth Hirth has surveyed the Amatzinac drainage of
eastern Morelos and the Amacusac and Chalma drainages of western
Morelos (Hirth 1974, 1976). Surveys and reconnaissances have also been
carried out in Puebla and Tlaxcala.

To illustrate the low intensity of surface collection in regional
surveys in central Mexico, it is instructive to note how many sherds per
site or component were actually collected by these regional surveys.
(Sites ranged in size from less than a hectare to over 50 hectares).
Blanton collected 10,000 sherds in his Ixtapalapa Peninsula Region
Survey in the Basin of Mexico. Since he had 227 components (the sum
of the total number of time periods represented at each site), he
collected an average of 44 sherds per component. For Sanders' 1126
Classic period sites in the Teotihaucán Valley, Charles Kolb (1973)
states that 8,624 rim and decorated sherds were collected, which is
69 sherds per Classic period site. Spores' (1972) Mixteca Alta
Survey collected 24,164 sherds from 124 sites, which is about 200
sherds per site. The number per component is not available. Hirth
(1974), in his Amatzinac drainage survey, collected a record 150,000
sherds, which is 480 sherds per component. This was due to the fact
that, since the ceramic sequence in the area was not well known,
laboratory study was required for periodization of the sites (Hirth,
personal communication).

At the opposite end of the spectrum is what can be termed site-
intensive survey, which involves collecting everything within certain
specified units such as grid squares or visible structures. This
technique produces information on internal structure of a site, and is
probably the only practical way to quantitatively assess distribution
of functional types throughout a large site. This technique,
using grid squares, was employed by Elizabeth Brumfiel (1976) at
Huexotla, a Late Postclassic site in the Basin of Mexico; by Paul R.
Tolstoy and Suzanne K. Fish (1975) at Coapexco, an Early Formative
site near Amecameca; and by Lewarch and Mason (1977) in the Coatlán
del Río area of western Morelos at several Late Postclassic sites.
As an example of intensity of collection employing this technique,
90,000 sherds were collected from a 2.8-hectare area (which is about

32,000 per hectare) at the site of Coatlán Viejo. Collections using
visible structures were made by the Coxcatlán Project at the Late
Postclassic site of Venta Salada in Tehuacán Valley (Sisson 1973) and
by the Teotihuacán Mapping Project at Teotihuacán (Cowgill 1974).

Although it might be assumed that the Monte Albán Survey would
fall into this category, this was not the case. The collection
technique employed at Monte Albán was that of regional survey. Survey
and collection units comprised 2006 ancient residential terraces
distributed over 6.5 hectares on the slopes of Monte Albán and other
nearby hills (Blanton 1978:7, 30). Each terrace was treated as a
site as in regional survey, and a grab sample of about 50 temporally
diagnostic sherds was taken from each terrace. Although visual
estimates of sherd density by phase were made in the field, these
were not employed in the final report (Blanton 1978) because the
field crews were not sufficiently trained in the recognition of
diagnostic types. Also, diagnostic types suitable for use with surface
material were not completely identified until after completion of
fieldwork (Kowalewski et al. 1978). Thus, all periodization was
based only on the sherds actually collected. This is more objective
than field estimation, but the small collection sizes and the grab-
sample nature of the 1700 collections greatly reduces the reliability
of the data. A total of 55,000 sherds collected were typeable, but
only 25,000 were diagnostic of single phases (Blanton 1978:11). Thus,
an average of only 15 diagnostic sherds were available from each
collection unit, which was usually a terrace (although more than one
collection was sometimes made at large terraces). If all six phases
were potentially present in a collection unit, they would be diagnosed
by less than three sherds per phase. (In actuality, this average was
increased somewhat by the many single-phase III-B/IV collection units.)

Although Blanton (1978:16) acknowledges that the Monte Albán
Survey collected fewer sherds per unit than the Teotihuacán Mapping
Project, in comparing the two projects he ignores the fact that there
was a qualitative difference in collection tactics, rather than just
a quantitative difference. Whereas Blanton's collections were grab
samples (the most obvious rim and decorated sherds were collected
until one had a bagful) (Blanton 1978:26), at Teotihuacán "all rim
and feature sherds, obsidian, and lithic artifacts and figurines that
they saw" were collected by the field crews from each collection unit
or "tract" (Cowgill 1974:369). The crucial word here is "all." If
all artifacts are collected, then the quantity of various classes of

artifacts from different collection units can be statistically compared, and quantitative statements about the distribution of artifact densities across the site or area collected can be made. Cowgill's maps of number of sherds per hectare at Teotihuacán and Blanton's similar maps for Monte Albán are based on fundamentally different kinds of data. At Teotihuacán, collections are statistically comparable; at Monte Albán, they are not.

In choosing a collection technique for the zone around the Xoxocotlán irrigation canal, a compromise was sought between the small-grab-sample technique of regional survey and the time-consuming technique of collection in small grid squares used in site-intensive surveys. A technique was sought which would maintain comparability of collections but could be carried out rapidly in the field.

Comparability between density figures from all collection units is approached by collecting all parts of all collection units in the same way. The best way to insure comparability is to collect all members of the category of artifacts one wishes to study from every collection unit (naturally, absolute comparability is a physical impossibility, as the chapter on potential biases will suggest). For the Xoxocotlán Survey, it was decided to collect all rim sherds visible on the ground surface, since this artifact category gives the most temporal information per artifact. Restricting collection to rim sherds increased collection efficiency, since they are easily recognizable and all other artifacts could be ignored. Thus, rim sherds contributed the largest amount of data relative to the problem of defining temporal changes in patterns of occupational density, while requiring a minimum of collection time in the field.

TACTICS

The actual collection technique employed was as follows. In plowed fields, crew members were spaced every third furrow (about three meters apart), a distance which approximates the field of vision of most people. By slowly progressing along the furrows, any crew member could perceive rim sherds in a field of vision that overlapped that of the adjacent fieldworker. In unplowed fields, a similar distance between collectors was maintained.

Rather than setting up a grid system, modern field boundaries (which in many cases correspond to prehistoric terraces) were used to delimit collection units (see fig. 4). This greatly reduced field

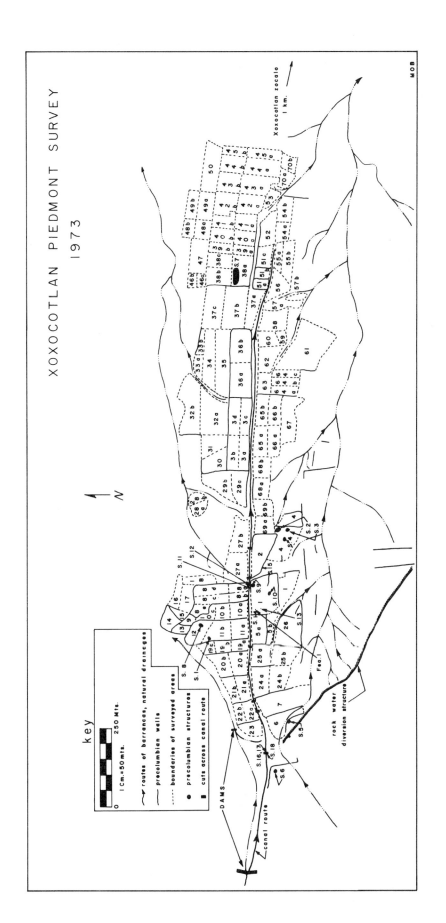

Fig. 4. Xoxocotlán Survey Project area showing collection units in relation to surface features

time, since collection units were already defined and were easily recognizable on aerial photos of the area. In some cases, where fields were large, they were arbitrarily subdivided and these boundaries were noted on the aerial photo. Most collection units varied in size between about 1,000 and 11,000 square meters (see table 15). One might argue that collection units were too large to adequately assess relative patterns of density in the 50-hectare zone around the canal. Judging by results, however, the size of collection units proved to be adequate for the problem at hand.

It is not maintained that every rim sherd from every collection unit was collected. It was expected only that rim sherds visible from standing height during one slow pass over the area would be collected. Thus, even though it is probable that not all rim sherds were collected, comparability was maintained, since each collection unit was similarly collected and all parts of all collection units were examined with equal intensity. Of course, perception of sherds varied from person to person according to time of day, type of vegetation, and other factors. These factors are impossible to control. However, field survey notes included a wide variety of information, such as type of soil, type of vegetation, percent of slope, and other pertinent data. Information on factors which differed between collection units can be used to determine whether there were consistent biases in the field crew's perception of sherds. Some attempts were made at minimizing effects of potential biases and controlling others. These efforts will be examined in detail in chapter XI.

DISCUSSION

The survey and collection technique described above may be termed "zonal" since it is intermediate in terms of time required per area covered between regional and site-intensive surveys. It combines the quantitative comparability of the collection units of site-intensive survey with rapid progress through an area of regional survey, since obvious topographic features visible on aerial photos, such as field boundaries, are used to delimit collection units. Speed is also increased if collection is limited to one obvious artifact category, such as rim sherds. However, it must be recognized that this practice limits the type of data retrieved and may not be suitable for problems involving functional interpretation, such as the definition of various kinds of activity areas. These problems probably require finer spatial control and retrieval of a wide variety of artifact categories including lithics, for which techniques of site-intensive survey and collection are more appropriate.

There is another sense in which the term "zonal" is appropriate for the technique described above. The result of this technique is a distribution of artifact densities across a zone. This precludes any necessity for having an a priori definition of a "site." One can define sites, or, in the case of this report, residential areas, quantitatively, on the basis of the pattern of densities found. This approach has been termed "siteless survey" (Dancey 1973, Dunnell and Dancey n.d.), since the goal is to plot the distribution of all densities relative to a zone, rather than merely to locate dense concentrations of artifacts or sites. The major difference between the two approaches is in the means of interpreting spatial relationships between units of analysis. Studies which emphasize locating dense concentrations use discrete coordinate information such that units of analysis are restricted to locations of "sites." In traditional regional survey, no information is imparted about the density of artifacts between sites. Thus, employing a zonal "siteless" strategy, one can discuss the distribution of artifact densities throughout the zone rather than just locations of sites (the unit of analysis shifts from the site to the zone). When the quantitative distribution of artifact densities has been plotted relative to the zone surveyed, one can infer the presence of sites, residential areas, activity areas, etc., based on quantitative data. The definition of a location as a site or activity area should not be taken as an a priori observable fact; instead, it should be seen as the first level of inference, which is based on data: the quantitative distribution of artifact densities.

3. The Late Formative Period

This chapter summarizes the development of a complex society in the Valley of Oaxaca, emphasizing the Late Formative period. This will provide a context for discussion of the Xoxocotlán data, which is mostly Late Formative. Knowledge of the Late Formative period is crucial to understanding the development of complex society in the Valley, since it is the period in which transition from relatively undifferentiated agricultural villages to a complex, stratified, hierarchically organized society (which probably integrated the entire Valley) occurred.

Before Flannery began the Oaxaca Human Ecology Project, it was thought that the Late Formative Monte Albán I phase was the earliest in the Valley and extended back to 800 B.C. (Bernal 1965). However, Flannery and his associates have defined a series of Early (1400-850 B.C.) and Middle (850-550 B.C.) Formative phases which precede Monte Albán I (Winter et al. 1975). The Early and Middle Formative periods are characterized by the growth of agricultural villages along the Atoyac River, making use of valley floor alluvium. By the Late Formative period there was a village located about every three kilometers along the river (Flannery and Schoenwetter 1970). Most of these Early and Middle Formative villages did not exhibit much internal status differentiation in residences and burials except at San José Mogoté, located in the Etla arm. San José Mogoté was approximately 10 times larger than the other villages, which averaged two to three hectares (Winter 1974a).

The Late Formative period in the Valley of Oaxaca saw major new developments in location and kinds of sites as well as an increase in status differentiation within sites. New sites were located in the piedmont zone, some of which seem to have been administrative centers located on hilltops overlooking perennial piedmont streams. These sites exhibit the first major use of stone architecture and are inferred to have exerted some control over irrigation water from the piedmont streams (Flannery and Schoenwetter 1970, Lees 1973). The only direct evidence for Late Formative irrigation, however, comes from Hierve el Agua, east of Mitla, and Xoxocotlán, below Monte Albán. Neither of these systems used perennial piedmont streams as their water source. The Hierve el Agua system (Neely 1967) had a spring as its water source, and the Xoxocotlán system depended on runoff impounded behind a dam.

Monte Albán was one of the new hilltop administrative centers with stone architecture, but it was not located near a piedmont stream. During the Late Formative period (550-150 B.C.) the site developed from three discrete residential areas around the main plaza area into a major ceremonial-administrative-residential center covering over four square kilometers (Blanton 1978) with impressive stone civic-ceremonial architecture and calendrical inscriptions.

Because of its size and central location, Monte Albán was un-doubtedly functioning as a regional center for much of the Valley by the end of the Late Formative period, but may have had competition from other sites such as Santa Maria Ayoquesco and Magdalena Apasco (fig. 1). Kirkby (1973) suggests that the territorial influence of Monte Albán may have been limited in the eastern Tlacolula arm by the presence of such sites as Mitla and Matatlán. As these two sites are located more than 10 kilometers from land which would have been agri-culturally productive during the Late Formative, they may have had the social or political control necessary to extract produce from land more than 10 kilometers away, while acting to control the flow of goods through the eastern entrance to the Valley.

The number of sites more than quadrupled between the Middle Formative and the beginning of the Late Formative in the Etla arm (Varner 1974). All Late Formative settlement types defined by Parsons (1971) for the Valley of Mexico (including primary and secondary regional centers, nucleated and dispersed villages, hamlets, and camps) have been found in the Valley of Oaxaca (Winter 1974a). Status differentiation within sites is also marked when compared with Early and Middle Formative sites. Winter (1974a) defines three types of residential units, which differ in degree of elaborateness, for the Late Formative, while in most Early and Middle Formative villages only one residential type is found. Burials also exhibit much greater status differentiation in comparison with Early and Middle Formative burials (Winter 1974a). This is especially evident at Monte Albán, where elaborate tombs occur.

Ceramically, the Late Formative period in the Valley of Oaxaca has been divided into the Early Monte Albán I phase (550-400 B.C.) and the Late Monte Albán I phase (400-150 B.C.) (Winter et al. ibid.). Some types previously thought to belong in Early Monte Albán I have been shifted to the recently defined Rosario phase (700-550 B.C.) of the Middle Formative period (Winter 1974b; Winter et al. ibid.). Early and Late Monte Albán I ceramics can be separated by a number of attributes, although several of them continue through both phases. One of the things which distinguishes Early Monte Albán I, however, is the presence of complex incised designs on the exterior below the rim. A similar Late Formative ceramic tradition, especially in terms of gray-ware, extends through the Late Cruz phase of the Mixteca Alta (Spores 1972) and the Late Santa María phase of the Tehuacán Valley (MacNeish, Peterson, and Flannery 1970).

The Terminal Formative or Protoclassic period is represented in the Valley of Oaxaca by the Monte Albán II phase (150 B.C.-A.D. 250). Monte Albán declined slightly in area during Monte Albán II (some peripheral areas were abandoned) and a large wall was constructed on the north and west sides of Monte Albán proper (Blanton 1978:54).

The Classic period is represented by the Monte Albán III-A phase (A.D. 250-450) and the Monte Albán III-B phase (A.D. 450-650). During Monte Albán III-A, Teotihuacán influence is present at Monte Albán and the site increases in area to about 4.7 square kilometers. The maximum extent of Monte Albán was reached during Monte Albán III-B (6.5 square kilometers) and Blanton's (1978:58) estimate of the population is 15,000 to 30,000.

The Main Plaza of Monte Albán was apparently abandoned after Monte Albán III-B but the continuation of Monte Albán III-B utilitarian ceramics into Monte Albán IV (A.D. 650-950) makes separation of the settlement patterns of these two phases difficult. The Monte Albán V phase (A.D. 950-1521) is characterized by a dispersed settlement pattern with many isolated residences.

4. Physical Environment

Detailed descriptions of the Valley of Oaxaca, its geology, environment, and physiographic features, can be found in Anne Kirkby's (1973) monograph on land and water usage. What is presented here are the main variables and conditions in condensed form for the Xoxocotlán piedmont area.

The area under investigation lies at the base of Monte Albán in the piedmont zone, which extends southeast toward the town of Xoxocotlán, located at the north end of the southern (Zaachila) arm of the Valley (fig. 1). Xoxocotlán lies approximately two kilometers south of the city of Oaxaca in the high alluvial zone, one kilometer west of the Atoyac River. The Xoxocotlán piedmont lies to the west of the town, rising gradually toward the mountain of Monte Albán to the west, and to the north toward a spur of Monte Albán, Monte Albán Chico. The land continues to rise irregularly to the west of Monte Albán all the way to the Sierras (plate 1).

The Xoxocotlán piedmont is dissected by deeply entrenched barrancas which have their heads on the steep slopes of Monte Albán. When water is funnelled into these barrancas during heavy cloudbursts, it represents a substantial erosional force. Between these barrancas are a series of farmable ridges. One of these generally northwest-southwest trending ridges, which begins directly below the South Platform on Monte Albán, comprises the survey area.

SOIL AND VEGETATION

Soil in this region grades from relatively infertile on the steep slopes of Monte Albán, where it is very rocky and extremely shallow, to very fertile nearer Xoxocotlán, where it has lost the characteristic reddish color of piedmont soil and taken on the characteristics of high-

alluvium soil--darker color and greater depth. The break between the
high alluvium and the low piedmont is very noticeable in some places
and not as noticeable in others. Just west of Xoxocotlán there is an
abrupt rise of two to 2.5 meters, which trends north-south for approximately a kilometer. This remnant may represent an old meander scar of
the Atoyac River, left behind as the river migrated eastward. This
line was taken to be the end of the high alluvium and the start of
the piedmont.

Native vegetation in the piedmont region near Xoxocotlán is
representative of that found in the piedmont zone elsewhere in the
Valley: mesquite, Opuntia (prickly-pear cactus) which may reach five
meters in height, a few scrub oak, and a variety of grasses. "Mala
mujer," a painful prickly nettle and the bane of almost every living
creature, inhabits rockier areas of the upper piedmont region, increasing in number as the elevation increases. In the barranca bottoms
in the low piedmont, cane occurs along with various broadleaf trees.
Where the cane is thickest, water is certain to be found during the
rainy season.

GEOLOGY

The soil in the upper piedmont is eroded in many places, exposing
limestone bedrock. Almost no other surface rock is present in the
lower piedmont--only the massive, nonfossiliferous limestone. In the
higher piedmont and lower mountain zone of the Monte Albán district,
limestone beds can be seen, separated from each other by differential
erosion. This exposure of limestone is especially evident on the
southeast side of Monte Albán above the area of investigation. Because of extensive slopewash, the outcrop becomes less distinct on
other sides of the mountain. The upper edge of the exposure is
relatively even, forming a generally horizontal band following the
contours of the slopes. This upper edge is characterized by a number
of unusually shaped whitish exposures, which may be travertine. In
the canyon to the north of the study area are a number of travertine
flows emerging from limestone beds. These flows are reminiscent of
the travertine flows found by Neely (1967, 1970) at Hierve el Agua.
There, and similarly in the Tehuacán Valley of Puebla (Woodbury and
Neely 1973), the travertine flows originated from spring heads; and
the waters, heavily laden with travertine particles in suspension and
solution, deposited these rock materials in and over irrigation features,

to preserve them through a process of "fossilization" or cementation.

The travertine indicates the presence of a satisfactory aquifer for
the water to pass through--perhaps only a series of deep cracks and
fissures. Travertine is nothing more than the deposition of calcium
carbonates and other rock materials resulting from an exchange or re-
placement of carbon dioxide for oxygen as the water breaks to the surface. But this is not a very common occurrence. Where it does occur,
it may help in the location of prehistoric canals, as at Hierve el
Agua or in the Tehuacán Valley. In the Xoxocotlán piedmont there are
no canals leading from this fossilized water source; it probably had
dried up long before the area was inhabited. Today, the travertine is
dry even during the rainy months. Two fossil springs, located downslope from the travertine slopes, were located, but these also were dry
year-round.

The location of the fossil springs on Monte Albán, perched some
150 meters above the Valley floor, is not geologically unusual. The
mountain range jutting up from the Valley floor, composed of Monte
Albán, El Gallo, and Atzompa, is a tectonic situation where strata
were faulted up to present the configuration seen today. Instead of
being the upthrown center of an anticlinal/synclinal structure, the
midvalley mountains are now seen to be erosional remnants of a
syncline. The original associated anticline has been eroded down on
both sides of the Monte Albán remnants to form the present valley.

The best cross-sectional view of the Valley formations is a few
kilometers to the north of Xoxocotlán, between the mountains of El
Gallo (which is actually part of Monte Albán) and Atzompa, along the
road to San Pedro Ixtlahuaca. On the south side of the road, one
can walk updip by traveling east and can pass through a series of
metamorphics, basal conglomerates, and sedimentary deposits of limestone and shale; after approximately one kilometer, one suddenly
finds oneself walking downdip, until the formations abruptly play out
when the edge of the present valley is reached.

The purpose of studying the geology was to determine whether or
not there might once have been a permanent source of water present which
would have supplied water throughout the dry season for the irrigation
system. However, the fact that Monte Albán is the erosional remnant of
a syncline means that there is no physical connection or aquifer for
water to travel along between beds of the northeast or western Sierras
and Monte Albán. The only water which would have been available
would have been runoff from rain which fell on Monte Albán and seeped

into the Monte Albán beds until an impervious layer was reached. This
water then would have been forced along the strata of Monte Albán, pre-
dominately downdip to the northeast, where wells have been excavated
by modern inhabitants of the lower north slope of Monte Albán Chico.
The gneiss in this area is severely fractured and fissured, allowing
water to percolate upward toward the surface. The perched fossil
springs on the east side of Monte Albán were probably another outlet
for this runoff, until recent erosional cuts breached the aquifer,
permitting the water to escape through seeps in the bottoms of barrancas.

MODERN AGRICULTURAL PRACTICES

During the Late Formative period this area was agriculturally
productive with the aid of canal irrigation based on runoff stored in
a reservoir. Today, with larger, more productive varieties of maize,
most of the piedmont zone is devoted to rainfall agriculture, depending
almost entirely on actual rainfall and slope runoff during the summer
rainy season. Sometimes fields are even located on low terraces within
arroyo courses and actually on the arroyo bottoms. This constitutes
quite a gamble, since an unusually hard rain well might wash away the
crop.

The area today is terraced to minimize slope erosion. Many of these
terrace walls are prehistoric and date back to the Late Formative
period; others are modern and are frequently rebuilt to minimize soil
loss. The general slope of the area is gentle enough in most places
to preclude the high, steep type of terracing (originally built to
support prehistoric residences) occurring further upslope on Monte
Albán. In many places, a single course of small rocks is sufficient,
whereas in others, three or four rows are stacked one on another to
provide a rough facing. Sticks and brush are also good barriers,
catching the soil as it erodes downslope and forming a dam which is
soon silted in.

Modern irrigation is present, although on an extremely small
scale. In a few areas small ditches catch slope runoff and divert it
onto fields located downslope. These ditches are no more than eight to
10 centimeters wide and extend for only 20 to 30 meters. The amount of
land they service is almost negligible. Also, a small earthen dam was
noted in one of the barrancas which functioned to divert water into
nearby fields. Local cultivators recalled the dam having been built
within the last 15 years. The dam measures two meters in height and

four meters in width. With these dimensions it can hold only enough
water for a few plots.

All arable land in the Xoxocotlán piedmont is cultivated, although
not simultaneously. It appears that at any one time, almost half of
the cultivable land is left fallow. Contour plowing of the soil is
practiced the majority of the time, with the traditional yunta (oxen
team), although tractors are making their appearance in the area.

The major crop raised in the Xoxocotlán area, as in the rest of the
valley, is maize. On most of the plots it is interspersed with beans
and squash. These three staple crops make up the bulk of domestic
intake, as well as the surplus crop for selling purposes. Maize is
also fed to the animals and can be easily banked for leaner times.
In some places, farmers seemed to be experimenting with peanuts,
since a third or half of some fields contained peanuts instead of
maize. In lower, wetter areas, small tomato plants were raised.

5. Archaeological Features on the Xoxocotlán Piedmont

The piedmont region between Xoxocotlán and Monte Albán contains an abundance of archaeological structures, some of which are obscured by the thorns and thick scrub brush which inhabit the upper reaches of this region. These structures and features include mounds, terrace walls, and irrigation facilities in use from the Formative period on. While the Xoxocotlán piedmont is not as densely covered with structures as are the higher slopes nearer the hilltop center, there is no paucity of remains. The survey revealed 18 surface mounds, two dams, a canal system, and an abundance of terrace walls (fig. 4). In addition, six other structures were found during excavation.

There is a definite break in the distribution of archaeological features as one progresses upslope toward Monte Albán. Near the bottom of the mountain, where the slope is steep, there are no structures present; it is only where the slope gradually diminishes at the start of the piedmont, that structures begin to appear. However, elsewhere at Monte Albán, especially to the north, where the slope is less steep and more conducive to terracing, structures were probably present at all levels.

HYDRAULIC SYSTEMS

It became apparent during the end of the 1971 field season that there had been quite an elaborate water-control system on the eastern piedmont of Monte Albán--one which impounded and then channeled water to various parts of the piedmont. The entire system consisted of two dams, one of which remained almost intact, the other of which was almost entirely eroded away, and two major canals. Of these two canals, the one exiting from the dam that remained intact was the more important, coursing its way downslope for over two kilometers before emptying into an arroyo.

The Main Dam/Canal System

Neely discovered the dam in the largest barranca in the Xoxocotlán piedmont during the 1971 field season (plate 2). It had gone unnoticed by Caso et al., due to its unassuming appearance and almost complete enclosure by brush. A modern trail winds along the southern end of the dam, but to actually see the structure for what it is, one must get down into the barranca in front and look back toward it. From there the construction, which appears in the face of an erosional breach through the dam, can be seen.

During 1972, the structure was cleared, and it was discovered that the dam was constructed of unmodified boulder fill into which a limestone mixture had been dumped as cement. A thick capping of this cement was then applied to the outer facing of the dam to seal it. In addition, the top meter of the dam was of cut, fist-sized limestone blocks, fitted together very neatly. The dam was not built as a rectangular structure spanning the barranca at right angles to the flow, but as a slightly V-shaped structure with the point of the V directed upstream. The gap in the middle of the dam probably contained a sluice gate for the regulation of water in the reservoir. In fact, in the southern face of the breach, the construction has such a finished appearance that it seems likely that this is an original side of the sluiceway which erosion has not destroyed. The northern face of the sluiceway has been severely eroded by water, which pours through after heavy rains. This erosion has formed a deep gully in the sediments deposited behind the dam.

The dam was approximately 10 meters high and 80 meters across the top when first constructed; the two ends tie in with two large limestone outcrops on either side of the barranca. From an engineering standpoint the construction seems to be ideal. At this point, the barranca is fairly narrow but still broad enough to allow for a fairly large reservoir. The exact dimensions of the reservoir were impossible to determine, but surface area was figured at about .6 hectare and maximum volume at about 18,000 cubic meters.

In the north canyon wall above the dam is a quarry which was probably the source for the limestone blocks used to dress the top of the structure. The rock was probably dislodged by means of wooden wedges, driven into seams in the outcrop which were then wetted,

causing the wedges to expand and crack the rock. Many fist-sized
blocks, similar to ones used in the dam construction, were lying loose
on the ground beneath the outcrop.

The major canal exits from the southern end of the dam and runs
along the contour of the mountain above the barranca until it reaches
the piedmont, where it runs out along a low ridge between two barrancas
(fig. 4). The canal is quite visible in some spots and less so in others.
On the higher slope it has cut down into bedrock in places and is extremely easy to trace. It is also visible when it reaches deeper soil
of the low piedmont. In between, there are spots where it can only
be inferred because severe slope erosion has destroyed any remains.
In two places, where the slope of the mountain is particularly severe,
masonry and rock were probably employed to provide a trough for the
water, although none of this remains today.

Once the canal reaches deeper soil, it is easy to trace, because
it has left a slight depression even though it has long been silted up.
On either side, paralleling the depression, are small lines of rock
rubble, probably the result of canal cleanings. The canal was originally
built in two levels: a smaller channel was excavated into the floor
and along one edge of a larger channel. Relative to the surface
indications of the canal and the paralleling spoil banks, the actual
channels of the canal are rather small. The larger, upper channel
averages 80 centimeters in width and 25 centimeters in depth, while
the smaller, lower channel averages about 30 centimeters in width and
12 centimeters in depth. In comparing the two channels it was found
that the larger channel, which, when functioning, included the lower
channel, had a carrying capacity 6.5 times greater than the smaller
channel.

The nature of the fill within these channels strongly suggests
that they were used contemporaneously. This would provide a very
efficient means by which two different (and relatively predictable)
flows of water could be distributed through the system. These two
different flows of water could result from controlled allocations of
water by the agriculturalists, but more likely resulted from differences
occurring in available water supplies between the winter dry season
and the summer wet season. The use of a smaller channel within the
larger is an efficient means of distributing smaller quantities of
water. By reducing both surface and subsurface area of the channel,
water loss through evaporation and seepage would be cut to a minimum.

Although no takeoff canals were found branching laterally from the

main channel, these must have existed. Terraced fields extend out to
the north and south and would have had to have feeder canals to service
them.

As the canal progresses downslope toward Xoxocotlán, it leaves the
ridgetop a little less than two kilometers from the dam. The canal
then runs along the south flank of the ridge below what turned out to
be the densest zone of occupation, which is located on the ridgetop at
this point. Beyond about two kilometers from the dam, a point is
reached where the canal can no longer be traced. The head of an arroyo
lies just below this point, but it appears to be the result of recent
erosion. It is possible that this arroyo follows the old canal outlet
route to the barranca which bounds the south side of the ridge.

Other Systems

Two other canals, upslope from the main canal, were cut parallel
to the mountainside to collect rain runoff and channel it into the main
canal. In reality, these canals probably also functioned to prevent
runoff water and soils from inundating the upper terraced fields and
thereby damaging the crops. These canals were constructed by utilizing
the natural slopeface of the mountain for one of the canal walls, while
the other was formed merely by building up a wall of earth and rubble.
The form of construction, the intermittent but forceful nature of the
runoff, and the accumulation of soils brought down with the water
runoff, strongly suggest that constant maintenance was necessary to
insure the survival and usefulness of these canals.

In the large barranca containing the dam described above (but further downslope from it) are the remnants of another dam. The only
remaining elements of this structure are a few large rocks poorly
cemented to a limestone outcrop in the northern side of the arroyo. The
feature which made the dam area easy to spot was the large amount of silt
deposited behind what once was the dam. This deposit was extremely thick
and extended back up the arroyo for approximately 30 meters, forming a
smooth, flat surface. The arroyo in this area is deeply incised and
very narrow. The wide, main canyon, of which this arroyo is the
eastern extension, plays out as it reaches the lower slope of Monte
Albán and this arroyo begins to deeply entrench itself into the
softer caprock of the piedmont. Therefore, the dam needed to have
been only 15-20 meters long when constructed. No canals were found
in association with this dam. It is not known if it was part of the

system described above or if it dates from some other period.

AGRICULTURAL FEATURES

The only Prehispanic agricultural features present (other than
the hydraulic systems) are terrace walls, constructed to retard soil
erosion and loss of nutrients through leaching. The terraces were
formed through the construction of from one to three walls of unmodified cobbles and boulders. While most of these walls appear to have
been dry-laid, there are some instances where mortar was used in the
construction.

There is a problem in trying to determine which terrace walls
are Prehispanic and which are more modern. Ideally, excavation would
be undertaken on each wall to determine its age. It is possible to
distinguish two types of walls, however: those in which loose rocks
were simply piled up in a long row and which were still above the
surface of the ground, and those where only the uppermost rocks were
evident above surface. These latter walls were usually fairly obvious, as there would be an accompanying drop in ground level in
front of them. They served as true terrace walls and soil-erosion
retardants, whereas the more modern-looking walls were used more as
boundary markers. In some instances, the presumed Prehispanic wall
is still effective in retarding erosion and creating level land upon
which to plant. Presumed Prehispanic terrace walls are shown as
solid lines in fig. 4.

Two walls were of special interest, as they reflect a great
deal of labor and ingenuity. One is an extremely large retaining wall
for an agricultural field (plate 11; in fig. 4, the wall bounds the
east side of Area 7). The actual slope of the land is not great,
thus the wall was built at an angle rather than vertically. It runs
for approximately 200 meters, from the canal on the north to a barranca
on the south. Only one and two stones thick, the wall functions more
as a protective capping over the soil than as an actual wall. The
flat area created as a result of this large retaining wall is
about 1.5 hectares, the largest artificially constructed field in the
area.

The other wall which should be mentioned is a terrace wall
surrounding Structure 9 (plate 3). This wall creates a terrace
which is over a meter higher than the surrounding terrain. To the
west, the ground gently rises to the same level as the field, and the

terrace wall ends. Construction of the terrace consisted of piling
up a meter of uncemented rock around the northern and eastern ends of
the field in which Structure 9 is situated, and then filling behind it
to raise the level of the field. As a result, part of Structure 9 is
below the surface of the terrace (see chapter VI for a description of
excavation of these features).

Agricultural terraces are not unique in this area. The lower
slopes of Monte Albán, especially on its north side, are literally
covered with these terraces, some ranging up to 300 meters or more in
length. Terracing is present (although not prevalent) in other parts
of the Valley of Oaxaca. Outside the Valley, especially to the
north where valleys are relatively narrow and steep-sided, terracing
is also necessary.

STRUCTURES

This section presents only the physical description of each
structure; details of excavation, ceramics recovered, burials, etc.
will be found in subsequent chapters.* A note on periodization of
each structure is given, although, for the excavated structures,
a more detailed report will be found in the chapter on excavations.
The locations of these structures are illustrated in figures 4 and 5.
Periodization of each structure, based on ceramic percentages computed
from surface collection, is given in table 1.

Structure 1

Description:

This low mound was tentatively given a structure number early
in the season, but had it not been for later reinvestigation, its
validity as a structure would have remained in question. What many times
seem to be mounds actually turn out to be limestone outcrops disguised
by a thin capping of dirt. By scraping the surface of Structure 1, a

---

* Since some structures were found during survey and others during
excavation, the structures could have been divided into the two
categories and included in different chapters. It was decided, however,
first to present all the structures together, giving their physical
descriptions, and in another chapter to discuss the excavations. This
enables the reader to skip chapter VI (on excavation) and still have
the basic information on all the structures.

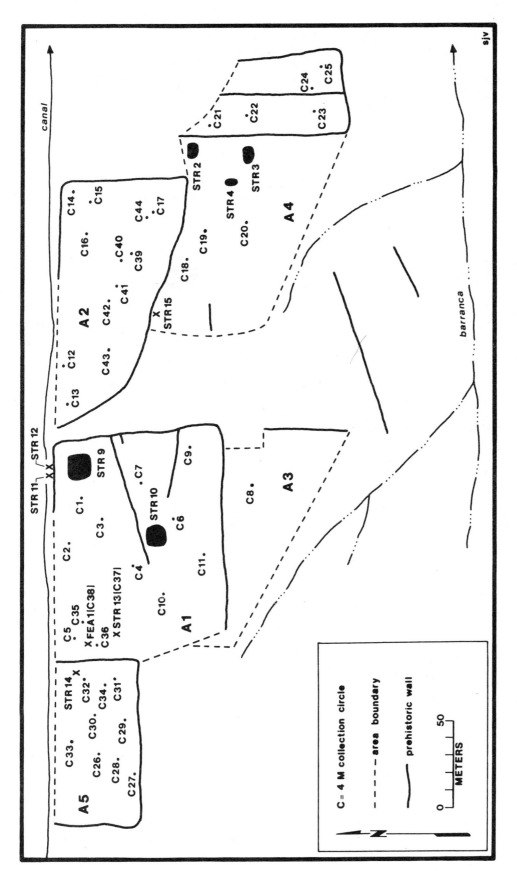

Fig. 5. Map of Areas 1-5, with associated structures, terrace walls, and four-meter-sample collection circles

wall was found loosely cemented to the bedrock. The size of the structure remains enigmatic, although it was quite small; the remaining mound is approximately seven meters square and one-half meter high. These dimensions are only estimates, due to the severely eroded condition of the mound.

Periodization:

The structure is situated in Area 11b, which has a heavy concentration of Early and Late Monte Albán I pottery plus fair representation of sherds from later phases. The pottery from Late I clustered in the northwest corner of the area, around the mound, which may be an indication of the age of the structure.

Structure 2

Description:

Structure 2 is one of a group of three mounds located in the western sector of Area 4. The mound is approximately 12 meters long, six meters wide, and one meter high. There is no visible evidence of construction. This structure, along with Structures 3 and 4, forms a U-shaped plaza with the open side toward the east. This arrangement is fairly common on Monte Albán: three mounds of unequal size arranged around a small plaza or patio which contains a centrally located adoratorio. Evidence of such a structure is indicated to the south of Structure 2 by a small, low mound two meters square (not given a structure number).

Periodization:

Identifiable sherds from the structure were all Late Monte Albán I (table 1).

Table 1. Periodization of rim sherds from surfaces of Structures 1-10

Structure	Early I	Late I	II	III-A	III-B/IV		Unknown
1	1	10	1		1	4	3
2		9					2
3		3	12				6
4	1	16					9
5		2			1	7	2
6		4				7	
7	2	27	4				15
8		5					1
9	8	79	9			2	27
10	1	24	6		1	4	16

Structure 3

Description:

Structure 3 is the largest of the three mounds surrounding the U-shaped plaza, and lies 20 meters south of Structure 2. Its dimensions are 15 by 10 meters, and 80 centimeters in height. Heavy erosion has taken its toll here, as it has on the other two mounds. No evidence of construction is visible.

Periodization:

Three Late Monte Albán I sherds were recovered, along with 12 Monte Albán II sherds, indicating that the structure probably was in use during both phases.

Structure 4

Description:

This structure is the smallest of the group and lies just west of Structures 2 and 3. No construction is evident, as a result of almost complete erosion. Its dimensions are approximately five meters on a side and one-half meter in height.

Periodization:

Sherds recovered date decisively from Late Monte Albán I.

Structure 5

Description:

Structure 5 is a mound one-half meter high, located adjacent to Area 6. Erosion has severely leveled the mound, obscuring its true dimensions. The best estimate of size is 10 meters square, although this is admittedly a guess. The mound is situated on a small terrace which projects out from the hillside, bolstered on three sides by a terrace wall one-half meter high.

Periodization:

Occupation may have been during Monte Albán V, as evidenced by the sherd count (although this is based on seven sherds out of a total of 12 collected).

Structure 6

Description:

Like Structure 5, Structure 6 is a low, severely eroded
mound situated on a small, artificially created terrace. The dimensions of this mound are also hard to judge, but it is approximately
eight meters square.

Periodization:

Occupation of this structure was apparently during Late
Monte Albán I and Monte Albán V (table 1).

Structure 7

Description:

Located in Area 38a, this structure is, to say the least,
very distinctive, being 40 meters long, seven meters wide, and one-
half meter high. This is the largest surface structure in the survey
area, but due to the shortness of the excavation period, no testing
of this structure was possible.

Periodization:

Sherd concentrations were fairly high in this general area;
Area 40a to the east had the highest density of any area collected.
Sherds from Structure 7 dated mostly Late Monte Albán I, although
sherds from Early Monte Albán I and Monte Albán II were present (table
1).

Structure 8

Description:

Structure 8 is a low mound located in the eastern half of
Area 12. The only construction evident is a little plaster eroding
out of one side. The sides of the mound are approximately seven by
eight meters, and the height is one-half meter.

Periodization:

Sherds from Area 12 around this structure were Late Monte
Albán I.

Structure 9

Description:

        Structure 9 is a 12-meter-square, two-tiered pyramid, rising 1.10 meters above the present land surface. The structure at one time was over a meter higher and had a third tier, but due to land-filling around its perimeter, the height has been decreased and the third (lowest) tier buried. The whole of the exterior was capped with plaster, little of which now remains (plate 7). Excavation of the structure revealed that it had been enlarged at least once. The inner construction was of large cut limestone blocks covered by hard-packed earth (plate 5). The secondary construction phase consisted of smaller, more irregularly placed stones arranged to form the two upper tiers. A small hearth was found on the top of the structure (plate 6).

Periodization:

        Sherds indicated that the structure was built during Late Monte Albán I.

Structure 10

Description:

        This seven-meter-square mound is located 30 meters southwest of Structure 9. It is approximately 1.5 meters high and presents no surface evidence of construction.

Periodization:

        No excavation was done on this mound; periodization must therefore come from sherds collected from the surface. A large proportion of the sherds were from Late Monte Albán I, with a few dating to Monte Albán II and V.

Structure 11

Description:

        Structure 11 was found during excavation through the terrace wall surrounding Structure 9 (plate 3). As the cut reached the base of the wall, it became obvious that it had revealed another structure situated well below the terrace wall. The floor of the structure is

composed of close-fitting flat flagstones bounded on the west by a
wall of well-cut square stones (plate 4). Both the floor and the
wall protruded past the present dirt embankment forming the perimeter
of the raised terrace, and were truncated by erosion. No trace of
the wall was found on the east side of the flagging.

Periodization:

Sherds were scattered throughout the fill above the floor,
but there was no primary deposit on the floor. Sherds recovered from
the fill date to Early and Late Monte Albán I.

Structure 12

Description:

This structure was also found during sectioning of the terrace
wall around the field containing Structure 9. It lies three meters
east of Structure 11 on the same ground level, and was at first
thought to be part of Structure 11. The wall which was exposed,
however, was at a 45° angle to the wall of the adjacent structure and
was composed of very angular rocks, unlike the hewn stone of Structure
11.

Periodization:

No sherds were found in direct association with the wall;
the sherds recovered from the fill above the structure dated from
Early and Late Monte Albán I.

Structure 13

Description:

Structure 13, which appeared as a slight rise above the
present ground, was uncovered by excavation. It lies 50 meters
west of Structure 9, and from the three meters which were opened, the
structure appears to be a house floor composed of angular chunks of
limestone. No walls were exposed, and it is doubtful that they even
remain, due to the shallowness of the soil in the immediate area.
Directly west of one edge of the floor or platform was a one-meter-
diameter circle of rock, the remains of a hearth.

Periodization:

Pottery was scattered throughout the fill above the rocks, but became more scarce toward the top of the rock layer. The condition and size of the sherds indicated that they were not part of a primary deposit. The surface sherds were a mixture of Early and Late Monte Albán I.

Structure 14

Description:

Located approximately 70 meters west of Structure 9, this structure was located solely by excavation. It appears to be the remains of a house, possibly with four rooms located around a central patio (plate 10). Due to the shallowness of the soil and modern plowing, no evidence of walls was found. The floor of the structure was composed of small rocks with a capping of hard-packed, burnt earth. The walls of the structure had collapsed inward, strewing larger rocks over this surface. Two individuals had been interred in the central patio area.

Periodization:

Sherds recovered from within the structure date from Early and Late Monte Albán I.

Structure 15

Description:

This structure was found while sectioning across what appeared to be a takeoff canal from the main east-west channel. All that remains of the structure is a single row of large stones, probably a building wall (plate 12).

Periodization:

Sherds excavated from the west side of it were decisively Late Monte Albán I and Monte Albán II, with the latter predominating.

Structures 16-18

Description:

Structures 16-18 consist of the remnants of three super-

imposed buildings located midway between the main dam and Structure 9,
The three construction phases were difficult to discern, but all
three apparently predated the canal. This became evident as the canal
cut was widened, and the canal was seen to be stratigraphically well
above the top of Structure 16, the uppermost construction,

Periodization:

Sherds from Structures 16 and 17, the upper two structures,
contained a few Early Monte Albán I rim sherds, but Late Monte Albán
I sherds predominated. Structure 18, the lowest structure, had rim
sherds which were split almost equally between the two phases.

6. Excavation

Since excavation had to be undertaken before the survey, there was little information upon which to base the decision of where to excavate. Excavation was undertaken in and around Structure 9 for a number of reasons. O'Brien was familiar with the area as a result of his work the previous year with Neely, when sherds were collected from the surface of structure 9 which indicated probable contemporaneity with the canal. Structure 9 is also the surface feature closest to the canal. Thus, excavations were carried out to test this association of mound and canal. These excavations produced Structures 9, 9-sub-1, 11, and 12.

Another objective of excavation was the exposure of living surfaces which would yield primary deposits of ceramic material to compare with that obtained from surface survey. In order to locate these sub-surface deposits around Structure 9, Areas 1, 2, 4, and 5 were sampled by means of four-meter-diameter circles. This is similar to the method used by Flannery at San José Mogoté (Flannery 1976). Since excavation had to be limited to two weeks at the beginning of the field season, there was no time to lay out a grid or devise a formal sampling strategy. The circles were positioned arbitrarily so as to cover all of the four areas fairly evenly, and some were placed where denser concentrations of sherds were obvious on the surface (fig. 5). Everything within the circles was collected and the top few centimeters were scraped with a trowel. The results of the sample-circle collections are given in tables 22, 23, and 24.

The four-meter-circle sample collections indicated that excavation in the upper or western end of Area 1 was most likely to produce a living surface. The excavation of circle 37 produced Structure 13, which is inferred to have been a house floor, and Circle 38 overlay Feature 1, a midden and cooking area. Structure 14, another probable house, was exposed near Circle 32, just across the terrace wall in Area 5.

All other subsurface structures were encountered during cuts intended to verify the presence of a canal. Structure 15 was discovered during the testing of a hypothesized takeoff canal (which turned out to be nonexistent). Structures 16, 17, and 18 were found upslope from the zone surface collected while sectioning the main canal route between the dam and the piedmont zone.

The fact that excavation had to be undertaken before the survey and was restricted to two weeks, was frustrating because entire structures could not be exposed, making interpretation difficult. If excavation could have been carried out after survey was completed, it would have been interesting to excavate around Structure 7, the largest structure in the area around the canal, which turned out to be in the most densely occupied part of the survey area. Excavations fulfilled their main purpose, however, in verifying that structures in the area were associated temporally with the canal. Additionally, excavations showed that the sherds from surface collections did reliably indicate the presence of coeval subsurface features.

STRUCTURE 9 AND STRUCTURE 9-SUB-1

As described in chapter V, Structure 9 is a 12-meter-square two-tiered pyramid rising 1.10 meters above ground level. The structure had been pitted at some time in the past, presumably by someone looking for a tomb. Erosion had partially filled the pit, so it was decided to clean up the west profile of the pit and extend it north to the edge of the structure (plate 3). This operation would minimize damage to the structure, while making construction technique visible and enabling the collection of sherds for dating purposes.

Trench 2 was begun in the field surrounding the structure (Area 1) and carried southward into the mound (fig. 6). As cleaning progressed, it became increasingly obvious that the pyramid had been constructed in two phases (fig. 7). There was a central core of large cut blocks with smaller rocks packed in, followed by a capping of hard-packed dirt. The soil matrix was a conglomeration of caliche and light-brown dirt with odd-sized chunks of mortar mixed in. Many of the larger rocks ranged up to one-half meter or more in length, but the average size was around 35 centimeters square. This inner construction was designated Structure 9-sub-1.

47

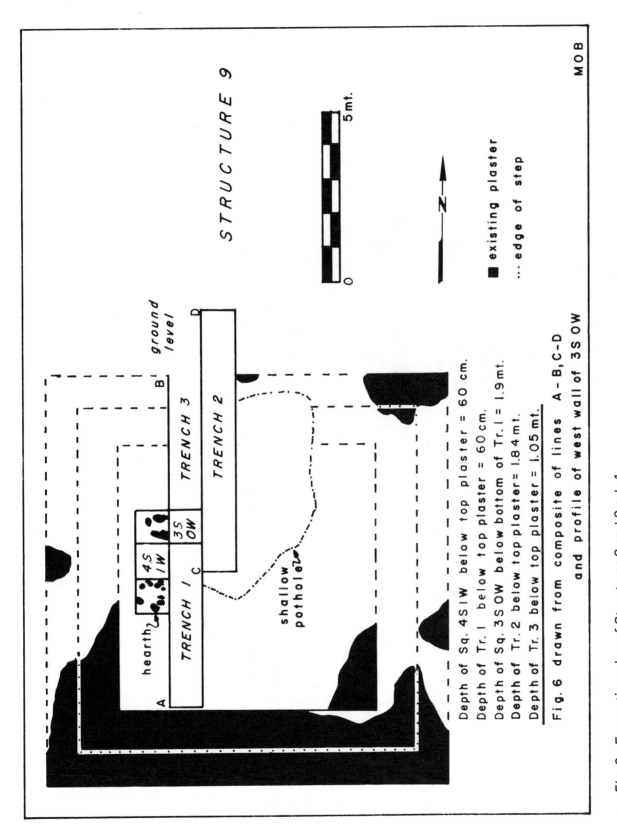

Depth of Sq. 4S1W below top plaster = 60 cm.
Depth of Tr. I below top plaster = 60 cm.
Depth of Sq. 3S OW below bottom of Tr. I = 1.9 mt.
Depth of Tr. 2 below top plaster = 1.84 mt.
Depth of Tr. 3 below top plaster = 1.05 mt.

Fig. 6 drawn from composite of lines A – B, C–D
and profile of west wall of 3S OW

Fig. 6. Excavation plan of Structures 9 and 9-sub-1

48

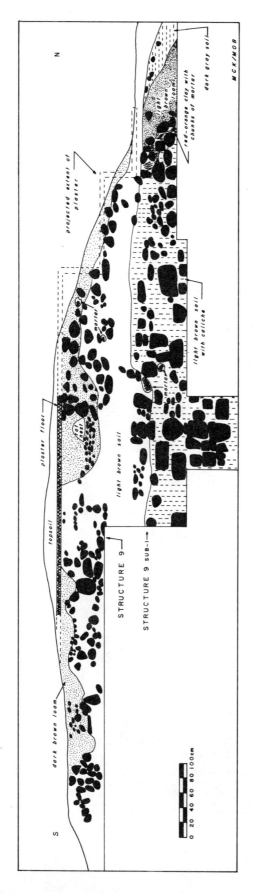

Fig. 7. Cross-section through Structures 9 and 9-sub-1

This substructure was built 12 meters on a side and shows no sign of ever having been plastered over. The height at time of construction was 1.4 meters (as evidenced by a later stratigraphic cut in Square 3S0W). Some well-cut stones were found in the northern end of the west profile, but further excavation was not undertaken to determine whether these were part of a short flight of steps to the top of the platform. This inner structure is entirely below the present field level and certainly antedates the filling-in of the land around the structure (figs. 7 and 8).

The secondary construction was exposed by Trench 2 (fig. 6), but to obtain a better profile the western face of the trench was stepped back a meter at the level of the base of the secondary construction (Trench 3). This trench was stepped up 45 centimeters to form Trench 1, an operation performed so that a lower rock layer of the upper construction could be mapped horizontally. Stones used for construction of the upper level (referred to hereafter as Structure 9) were much smaller than those used in Structure 9-sub-1, most of them falling under 20 centimeters square. In addition to a difference in size of the rocks used for the two structures, there was a distinct soil change between the construction levels. The hard-packed soil of the lower structure gave way to a much looser, brown soil which extended up through Structure 9.

Structure 9 rests on the dirt capping of Structure 9-sub-1 and was inset 95 centimeters on each side (fig. 8). This made the tier almost a 10-meter square. The uppermost tier was inset 1.2 meters on each side, forming a 7.6-meter square. The lower tier is 46 centimeters high and the upper one 59 centimeters high. They are the only two levels of the entire structure visible today above ground surface. Capped with 20 centimeters of dirt, the structure appears as a 1.25-meter-high mound.

After Structure 9 was completed, it was plastered over on the top and sides (plates 5 and 7). This included plastering over the dirt and rock surface of Structure 9-sub-1 where it showed around the edges of the lower inset tier of Structure 9. A primary coat of plaster was laid down, but later a single layer of rock was placed over the plaster on the two lower tiers, then another coat of plaster was laid down, raising the steps (or tiers) by the thickness of this rock layer.

In essence, the structure has three tiers, counting Structure 9-sub-1 as a buried tier of which only the surface is exposed above ground. In the few places where the plaster remained on the surface

50

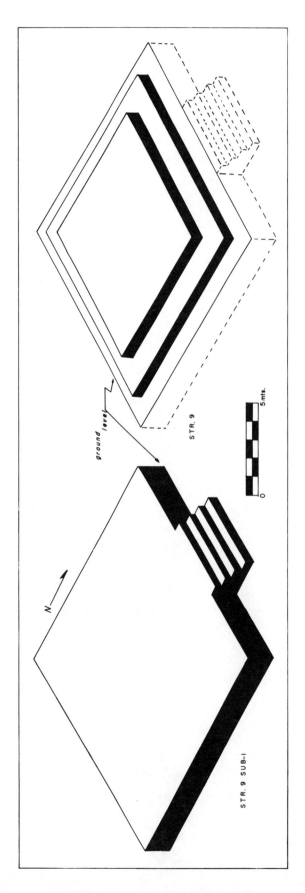

Fig. 8. Proposed reconstruction of Structures 9 and 9-sub-1

of the uppermost tier, it appeared burned (fig. 9). A small hearth
was found in Square 5S1W (fig. 9 and plate 6), but it could hardly
account for much of the structure being burned.

A burial was uncovered on the middle tier on the south side of the
structure (Burial 1; see chapter IX and plate 8). The body had been
placed at a right angle to the tier so that it extended down to the
next tier. A row of stones had been cemented parallel to the tier,
midway between the back of the tier and the front edge. This row of
small stones was 48 centimeters long, and formed the leading edge of
a small tomb as it projected out from the rear of the step (tier). The
body had been placed headfirst into the tomb, but the tomb covered no
more than half the torso--from the head to the middle of the back (plate
7). The top of the tiny enclosure had since collapsed, crushing the
cranium. No grave goods accompanied the body. The burial was probably
placed on the step during a later period and has no relation to the
early history of the mound.

For stratigraphic purposes, Square 4S1W was cut down through the
plaster floor to just above Structure 9-sub-1, a depth of 60 centimeters
(fig. 6). Square 3S0W was taken down to sterile soil. Rim sherds by
phase are given for each of the 20-centimeter excavation units in table
2.

Looking at the tabulation of sherds from the stratigraphic excavations (table 2), it is evident that both structures were built during
some part of Late Monte Albán I. There is no ceramic break between
Structure 9-sub-1 and Structure 9; the Early Monte Albán I sherds
present in both structures are there as a result of being in the fill
used for construction. On examining the sherds, Winter stated that
it was his feeling that the seven Monte Albán II sherds were early
Monte Albán II and might even extend back into the Late Monte Albán I
phase. Six of the seven sherds occur well within the upper meter of
the mound, and suggest that it was well toward the end of the Late
Monte Albán I phase when Structure 9 was added. No phase later than
Monte Albán II is represented among the identifiable sherds.

STRUCTURE 11

Lying nine meters north of Structure 9, Structure 11 was found
during excavation through the terrace wall around Area 1 (fig. 10).
The cut was made primarily to date the terrace wall in relation to
Structure 9, but as the trench reached a depth of one meter below

52

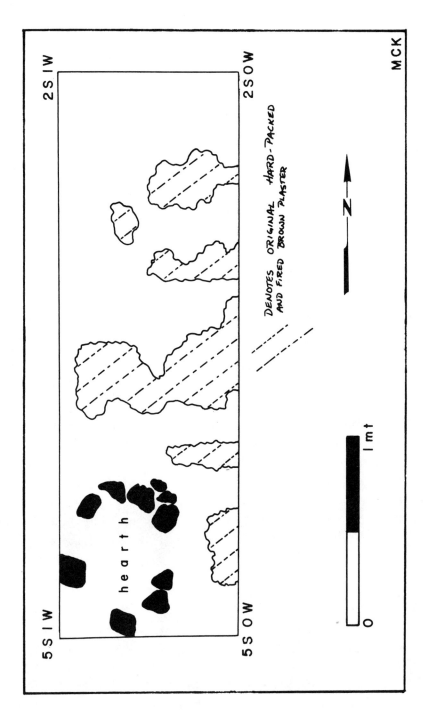

9. Plan of plaster floor and hearth, Structure 9

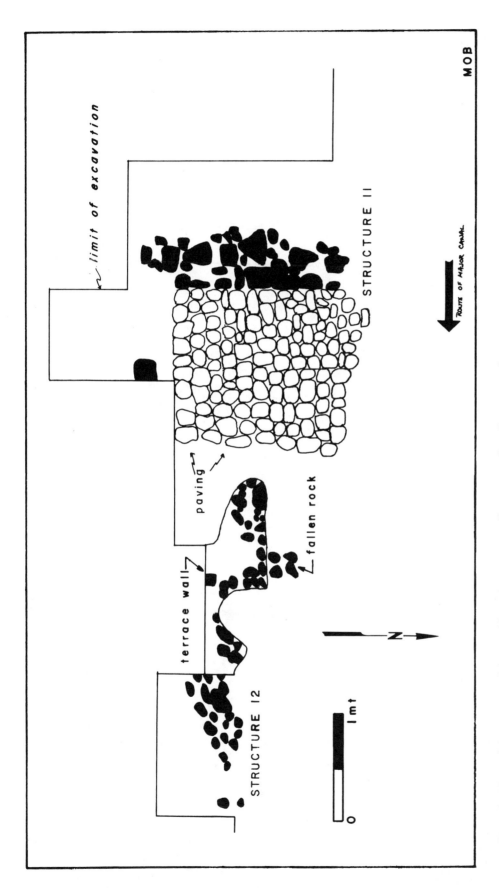

Fig. 10. Plan of Structures 11 and 12 and terrace wall surrounding Structure 9

the terrace wall, it became obvious that another structure had been
found. All that remains of this structure is about 2.5 square meters
of paving and a two-meter expanse of wall which is oriented 8° east of
north. The paving is of small, flat, uncemented rocks fitted very
closely together, bordered on the west by a wall 15 centimeters in
height composed of well-cut, square stones (plate 4). No wall was
found on the east side. Both the flooring and wall had at one time
extended out past the present edge of the raised field (area 1) and
accompanying retaining wall, but have since been truncated by erosion.

The structure is on the same level as Structure 9-sub-1 and may
have been contemporaneous with it. It appears that Structure 11 may
have been a small house; what is left is the remains of either a
patio or a room. Sherds were moderately scattered in the fill and
wash above the floor, but there was no primary deposit. Sherds
recovered were mostly Late Monte Albán I, with a few scattered Early
I sherds.

Table 2. Periodization of all rim sherds by excavation level from Squares 4S1W and 3S0W of
Structures 9 and 9-sub-1 (Structure 9-sub-1 starts at Level 6, 100-120 cm)

Depth in cm below plaster floor	Early I	Late I	II	Unknown
0-20		19	3	4
20-40	2	19	1	3
40-60	5	22		1
60-80		21	1	
80-100		8	1	5
100-120	2	10		
120-140	3	38		6
140-160	2	11		2
160-180	2	9	1	
180-200	1	13		
200-220		6		

STRUCTURE 12

Structure 12 is located a little over two meters east of Structure
11 and extends back under the raised field containing Structure 9
(fig. 10). It consists of a single wall of indeterminable length, as
only a meter of it was exposed. At first it was thought that this
wall was part of Structure 11, but the wall is at a 43° angle to the
latter structure. In addition, the stones in the Structure 12 wall
are very angular and rough and are unlike the hewn stones in the
Structure 11 wall.

Structure 12 is on the same level as Structure 11, both lying one meter below the present terrace level and just to the south of the canal and the main Xoxocotlán-Monte Albán trail. It is impossible to say for sure which one of these structures is earlier or later than the others, or which, if either, is contemporaneous with Structure 9 to the south. One guess is that Structure 12 is older than both Structures 9-sub-1 and 11, which may be contemporaneous, but this is based solely on the fact that Structure 9-sub-1 and Structure 11 are aligned in the same direction, while Structure 12 deviates from this pattern by more than 40°. No sherds were recovered from around the wall; those in the one meter of fill above it were predominantly Late Monte Albán I, with some Early Monte Albán I.

STRUCTURE 13

Excavation of Structure 13 consisted of three squares taken down to a layer of fist-sized limestone blocks, which created a rough flagging (fig. 11). Surface collections had indicated a high density of sherds in this area, and it also appeared from the surface that a structure might be present because of a low rise and soil-color change.

Actually, there were two areas of flagging, as evidenced by fig. 11. The more easterly of the two is a rectangular area, both sides of which were located. The exact length of this structure was not determined, due to insufficient excavation; the width is 1.25 meters. Only the northeastern corner of the other area of flagging was exposed, and from that small amount it is impossible to say exactly what kind of structure is present. There may be two separate structures, or the two areas of flagstone may be separate rooms of one structure whose walls have been plowed up.

Pottery was moderately scattered throughout the fill above the structure floor but became more scarce toward the floor. The surface collection contained few diagnostic Early Monte Albán I sherds, and the subsurface sample directly mirrored surface findings (see tables 9 and 10, chapter VII).

AREA 1, FEATURE 1

Surface conditions (sherd concentration in the four-meter-circle samples and a slight rise in the ground surface) indicated a probable

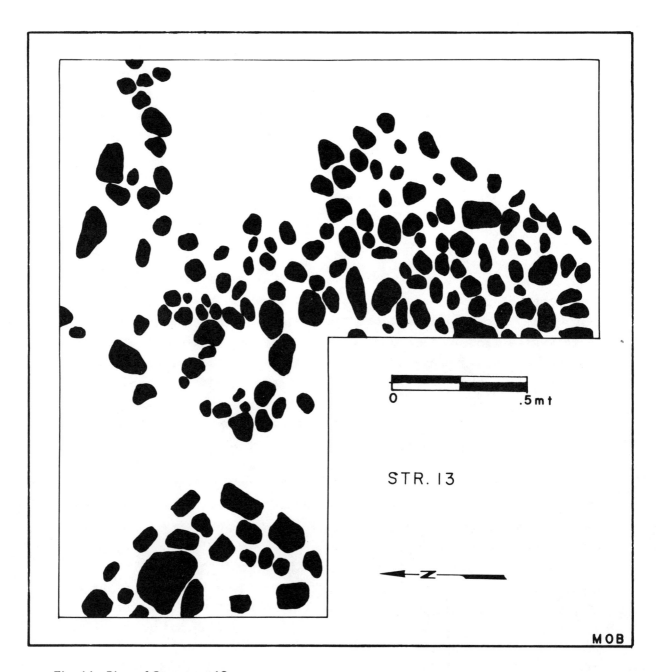

Fig. 11. Plan of Structure 13

subsurface structure here, but excavation revealed a midden area. The rise in ground surface was due to a rise in the bedrock which was covered by less than 30 centimeters of soil. Sixteen one-meter squares were excavated to bedrock in this area. In part of the area excavated, a depression had been chiseled out of the bedrock. This depression contained primary midden deposit with vessels broken in place. Three small pits were located within the larger depression, one of which (Pit 2) contained charcoal specks and gray earth (fig. 12, plate 2). The largest pit was 60 centimeters in width and 35 centimeters deep. Their small size and the presence of charcoal suggests that they were cooking pits or hearths, rather than storage pits.

It is possible that a house had been constructed directly on top of bedrock here and that the walls had since been plowed away, but it is more probable that a structure existed to one side of the excavated area and that the place functioned as a dump and cooking area. Winter (1974b, fig. 2) illustrates a similar situation from his excavation on Terrace 72 on Monte Albán. Structure 14 is eight meters west of Feature 1 and may be the primary structure associated with the bedrock midden area. Sherds from Feature 1 were from both Early and Late Monte Albán I, with Late I types predominating. Tables 6, 7, and 8, chapter VII, present the distribution of sherds by phase and type. Due to the fact that the plow zone extended down to bedrock in squares 2N1E, 1N4E, ON0E, 2S2E, 2S3E, and 2S4E, sherds collected from these squares are not figured in the totals for the feature.

STRUCTURE 14

In searching for the structure associated with Area 1, Feature 1, an area eight meters to the west was excavated. Fifteen to 20 centimeters below ground level, a layer of large, angular rock was exposed; upon expansion of the area, a zone of smaller rock was found to the west (fig. 13). A burnt clay overlying the rock layer was traced in two areas, and it seems probable that the burnt layer completely covered the sharp, angular rocks as a hard-packed floor. To the east, the rock layer abruptly stopped and skirted a square, 10-centimeter-deep depression, which may have been a central patio, although no plaster was found (plate 10). This central area measured 1.5 meters north-south, but no measurement is available for the east-west distance, due to lack of time for excavation to the east.

58

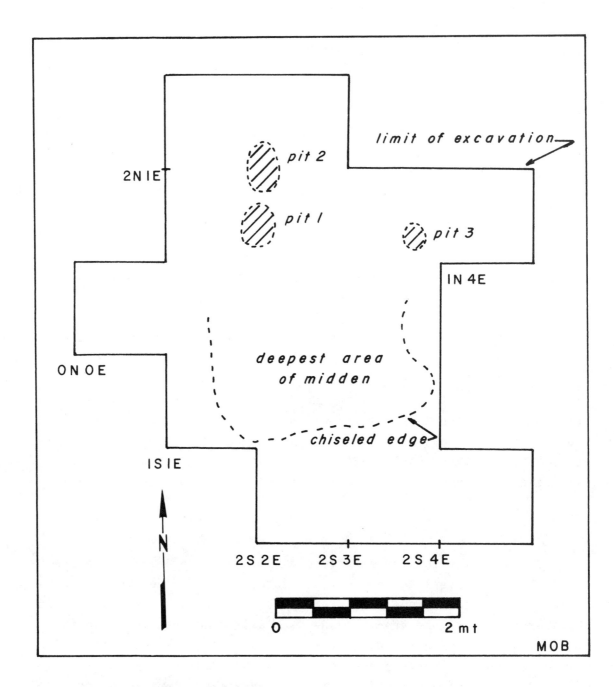

Fig. 12. Plan of Area 1, Feature 1

59

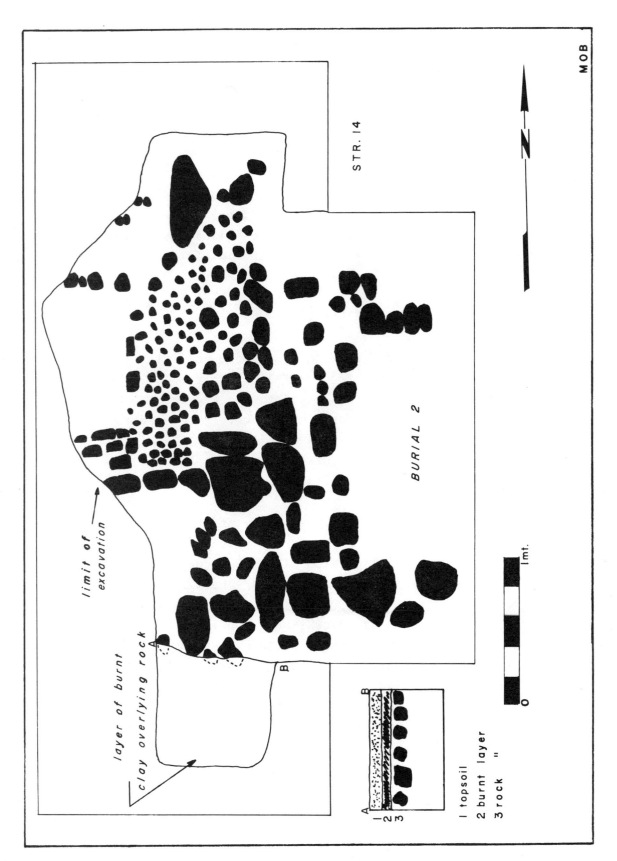

Fig. 13. Plan of Structure 14 and cross-section illustrating rock and overlying hard-packed clay

Burials 2 and 3 were found in the center of this open area.

This structure is probably the one associated with the midden
deposit to the east, although without excavation of the ground between
the two areas it is impossible to tell. Sherds were heavily concentrated above and on the rock layer and, where it remained, the burnt layer.
Many of the sherds were as large as those found in the deposit eight
meters to the east. The ceramics were entirely restricted to Early
and Late Monte Albán I phases. This is in keeping with what was
recovered from Feature 1 to the east. There was no internal stratigraphy to the structure, and thus no division between Early and Late
Monte Albán I strata was evident. (For a more detailed discussion
of the ceramics and a comparison with those found in Feature 1, see
chapter VII.)

STRUCTURE 15

Structure 15 consists of a four-meter-long rock wall with a few
associated flagstones on the east side (fig. 14 and plate 12). The
wall is oriented 8° east of north and contains what appears to be a
doorway on the west side. Eleven square meters were excavated to
various depths, three of which hit midden deposits (fig. 14). The
north end of the trench abutted the southern edge of a large terrace
wall which had eroded down over the bedrock shelf to the south (fig.
15).

Bedrock here is approximately .5 meter deep, and Structure 15
lies 10 centimeters above it. The structure is probably a house,
judging from the trash middens associated with it. However, the
midden deposit located in test pits 1 and 5 began about 10 centimeters below ground level and, hence, was higher than the structure
wall, the northern end of which is 40 centimeters below ground level.
This midden deposit may be from another structure, which was probably
located up on the terrace to the north. The cut through the eroded
section of the terrace wall (fig. 15) was too far east for the deposit to show up in profile, but the midden probably extended right
up to the base of the wall. Burials 4 and 5 were recovered from this
deposit. Sherds from all five test pits were Early and Late Monte
Albán I and Monte Albán II (table 3). There was no visible stratigraphy, and sherds from all three phases were mixed.

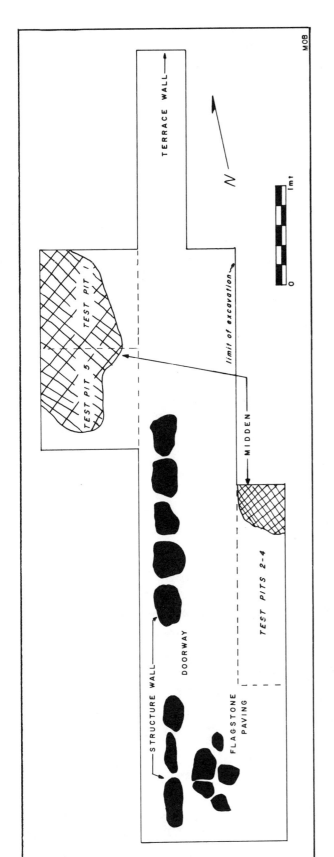

Fig. 14. Plan of Structure 15 and associated midden deposits

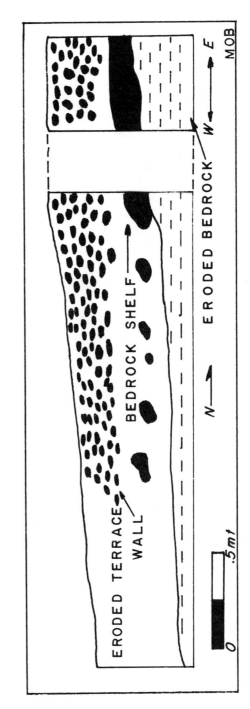

Fig. 15. Cross-section along west wall of trench containing Structures 16, 17, and 18

Table 3. Frequency of rim sherds from Structure 15 test pits by 20-cm levels

	Early I	Late I	II	III-A	III-B/IV	V	Unknown
TEST PIT 1							
0-20							2
20-40		14	13			5	
40-60	2	17	26				13
TEST PIT 2							
0-20			1				
20-40		2	5				2
TEST PIT 3							
0-20	1	2					3
20-40	1	11	4				5
TEST PIT 4							
0-20		2	1				5
20-40		4	2				3
40-60		5	10				14
TEST PIT 5							
0-20		2	1				6
20-40		2	5				8
40-60		19	27				21
60-80		7	11				6

STRUCTURE 16

In 1971, while sectioning the main canal halfway upslope between
Area 1 and the largest dam, Structure 16 was uncovered. What was
assumed to be the wall of a house or large platform appeared in the
east profile (figs. 16 and 17). Expansion of the area during 1972
showed that the feature was indeed a wall composed of limestone
blocks ranging in size from 10-20 centimeters up to 80 centimeters.
The wall was not the facing of a platform, but a free-standing
structure approximately 2.5 meters long and one-half meter wide.
It was aligned 4° east of north.

From the number of large sherds found in the fill surrounding
the wall (fig. 16, Zone 2), it was decided that the structure was
not a retaining wall for a terrace. Previous cross-sections through
terrace walls had shown that sherds did exist in the fill and matrix of
the walls, but not in the condition the sherds around Structure 16 were
in. Several partially reconstructible vessels came from the area.

Zone 2 (figs. 16 and 17) had the consistency of a midden deposit:
tannish-gray sandy soil with small flecks of gray ashlike material in
it. Zones 3 and 4, which appeared to be slopewash, were truncated
prior to construction of Structure 16. The two zones formed an
abutment for a probable east-west wall which ran perpendicular to the
existing one.

63

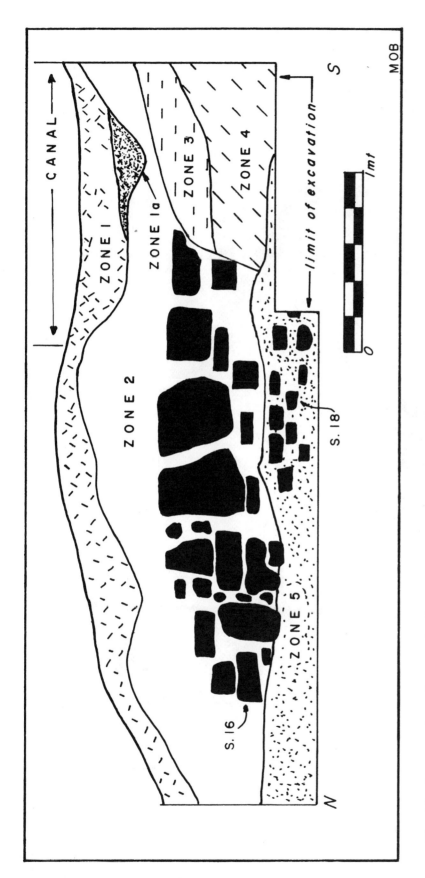

Fig. 16. Cross-section of east wall of trench containing Structures 16, 17, and 18

64

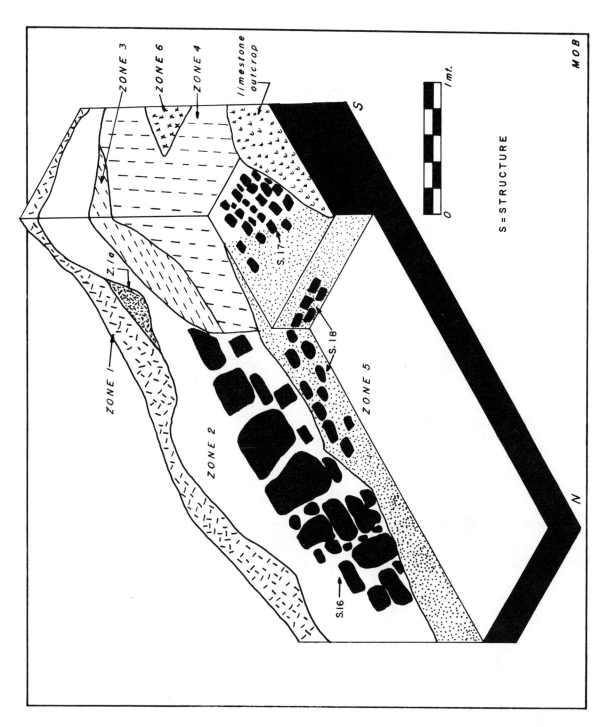

Fig. 17. Three-dimensional relationship of Structures 16, 17, and 18

The importance of this structure is in its stratigraphic position: directly beneath the canal (fig. 16). This structure, along with Structures 17 and 18, predates construction of the canal. Sherds from Structure 16 all date Early and Late Monte Albán I, with the latter outnumbering the former five to one (table 4).

Table 4. Distribution of rim sherds by phase for Structures 16-18

		Early I	Late I	Unknown
Structure	16	4	20	2
Structure	17	3	10	1
Structure	18a	7	8	
Structure	18b	6	8	1

a: from midden area around structure
b: from structure

STRUCTURE 17

The discovery of a one-meter-square area of flagging underneath Zone 4 (the lower layer of slopewash) led to the documentation of Structure 17. It was initially thought that this layer of flagging was associated with Structure 16. Upon examination of the profile, however, it was seen that this was a physical impossibility; Structure 17 had been built prior to the area's being covered by Zones 3 and 4, the slopewash (figs. 16 and 17). Structure 16 had been erected after this period of slopewash. Therefore, Structure 17, whatever it had been, predated Structure 16. Sherds taken from just above the level of flagging dated Early and Late Monte Albán I, but since that area was the slopewash zone, much mixing had occurred. However, since Structure 17 predated Structure 16, it had to date at least to some part of the Late I phase.

STRUCTURE 18

Structure 18 consists of a layer (platform?) of undetermined size, constructed of small limestone blocks 10-15 centimeters square. The structure lies in a midden zone stratigraphically distinct from the zone associated with Structure 16 (figs. 16 and 17). Without

66

this inconformity the two structures would be hard to distinguish
from one another. Due to lack of time, the dimensions of the structure could not be obtained. There is a possibility that the flagging
which constitutes Structure 17 could in some way be related to Structure 18, but this did not seem to be the case. Structure 17 sat
slightly above and to the south of Zone 5, the midden deposit associated
with Structure 18 (fig. 17). Ceramic material was collected in two
lots: sherds from the area above Structure 18 (but still within Zone
5) and sherds directly on top of the structure or visible within the
soil between the stones (table 4).

Except for Structure 9, the excavation of Structures 16, 17, and
18 was the only one which showed vertical stratigraphy with a superposition of structures. All of the structures seem to have been
constructed during Monte Albán I, probably mostly during Late Monte
Albán I; all predate the canal. This does not mean, however, that
there was no canal earlier. The canal route was probably moved
laterally from time to time, so that the canal exposed in profile
above Structure 16 was the location of the last canal before the
system was abandoned. This last canal is visible in cross-section
because it was abandoned and silted in. Earlier, the canal may have
run along the north or south sides of Structures 16, 17, and 18.
Construction of these structures probably destroyed evidence of this.

The superposition of these structures is of interest because it
indicates quite a bit of activity in the same place in a relatively
short period of time. One could speculate that these structures may
have had something to do with the canal and water control, since
they were directly on the canal route at the point where cultivable land
begins and were frequently rebuilt, perhaps every time the canal was
enlarged or relocated. Unfortunately, not enough area was opened up
to determine the function of these structures.

CANAL CUT 3

A total of six cuts were made across the canal to examine its
construction and to obtain sherds for dating purposes. Figure 18
illustrates a typical cross-section through the canal. Zone 5 is
a midden deposit of an unlocated structure into which the canal
(Zone 4) was cut. The canal is slightly less than a meter wide
here and has a secondary channel cut into the bottom. Zone 4 is

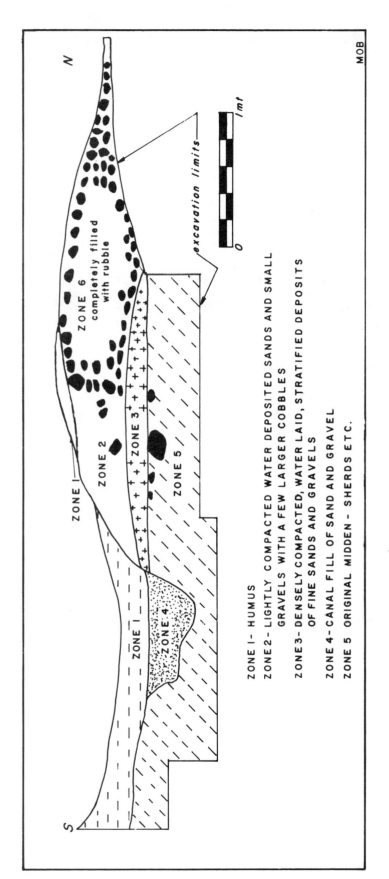

ZONE I – HUMUS

ZONE 2 – LIGHTLY COMPACTED WATER DEPOSITED SANDS AND SMALL
GRAVELS WITH A FEW LARGER COBBLES

ZONE 3 – DENSELY COMPACTED, WATER LAID, STRATIFIED DEPOSITS
OF FINE SANDS AND GRAVELS

ZONE 4 – CANAL FILL OF SAND AND GRAVEL

ZONE 5 ORIGINAL MIDDEN – SHERDS ETC.

Fig. 18. Canal cross-section excavated west of Area 7

a mixture of sand and gravel which washed down the canal and eventually
silted it in.

Zone 3 is a water-laid deposit of fine sand and gravel, deposited
as the canal overflowed during periods of heavy rain. This being the
natural downslope side, and slightly flatter than the other side
of the canal, it received this thin deposit, whereas the southern
side (fig. 18, left) did not.

Zone 2 is composed of coarser sand and gravel than appears in
Zone 3, and seems to be the result of excessive flooding or, more
probably, canal cleanings. This zone was laid down in front of a
large terrace wall (Zone 6) which was constructed directly on top
of Zone 3.

After canal abandonment, Zone 1 (a humic layer) was laid down.
Even though the canal proper is buried, the depression remains faintly
evident on the surface. Without this slumping the canals would be
difficult to locate.

Sherds recovered from the fill of the canal (zone 4) in this
and other canal cuts dated to Late Monte Albán I and Monte Albán II.

7. Ceramic Artifacts

Rim sherds from the surface collections made up a large percentage of the total sherd inventory. During excavation, material was screened through .5-inch mesh. This included material from Structures 9, 9-sub-1, 13, 14, 15, 16, 17, 18, and Feature 1. During this procedure all rim sherds were saved. In addition, all sherds from Pits 1-3 in Area 1, Feature 1 were recovered. After material was washed and sorted in the laboratory, vessel reconstructions were attempted. If body sherds were still unattached, they were bagged separately and set aside. Typing of sherds was done in Oaxaca, where there was access both to comparative material and to Winter, whose advice was of great assistance. When analysis was complete, sherds were turned over to the Instituto Nacional de Antropología e Historia, for storage in Cuilapan, Oaxaca.

Figurines represented only a fraction of the total amount of ceramic artifacts recovered, with only 33 coming from survey and seven from excavation.

POTTERY

Pottery is the most common artifact found on Mesoamerican sites, and this was especially true in the Xoxocotlán Survey, where it accounted for over 99% of material recovered from surface survey and excavation. Only one complete vessel was found during the entire work, and in very few instances was it possible to restore a vessel even partially. Small sherds make up almost the entire collection. For reasons discussed in chapter II, only rim sherds were collected during survey.

For each temporal phase there are good diagnostic types which are limited almost exclusively to that one period. There are other types, however, which last from one period to the next. As discussed in chapter II, to be able to delimit areas of growth on the Xoxocotlán piedmont, it was necessary first to divide the piedmont into small units and then to date the individual units. To do this, sherds were taken from each collection unit, and the rim sherds were sorted by period.

Classification and Valley of Oaxaca Ceramics

The problems surrounding ceramic classification in the Valley of Oaxaca have been only partially resolved. For the 3,000-year span from the Early Formative period to the Conquest there are periods for which the ceramics are well known, and there are those where the chronology suffers. Bernal (1947, 1949a,b) began preliminary typing of the Oaxacan sherds as Caso (1938) had done as early as the late 1930s. It was not until 1967, however, that a final description of sherds from almost 30 years' work at Monte Albán was published (Caso, Bernal, and Acosta 1967).

The pottery analysis undertaken by Caso, Bernal, and Acosta parallels that of Ford (1962) and others. It was based primarily on complete vessels removed from tombs, nonenclosed burials, and offerings. It has the advantage, then, of working with more than miscellaneous sherds. Each of their types <u>theoretically</u> consists of vessels characterized by a particular paste, surface treatment, decoration, and variety of shapes. It can also be said that each of these types occupies a specific time period and a definite geographical area.

Useful ceramic classifications are rare, since they must be refined enough to be of stratigraphic help but at the same time be fairly easy to use. It is of little aid if no two people can agree on where to place the object to be typed. The present Oaxacan ceramic classification can be criticized on both of these points. This is not to say that Caso, Bernal, and Acosta's work was not a monumental step forward; it is simply that the classification falls short of the ideal.

One of these obvious shortcomings is in defining what the range of variation is in Monte Albán III-A sherds. As of now, carved cylindrical types can easily be distinguished if the sherd is from part of the vessel which contains the carved design. A few other diagnostics occur from this phase, but on the whole, the number of recognizable Monte Albán III-A sherds in any surface collection is low. This paucity exists both in the Valley and on Monte Albán.

Another shortcoming of the typology is the way in which Monte Albán III-B and IV are divided on ceramic grounds. The two phases are difficult to distinguish, with the result that they are usually mixed together.

Caso et al. made a distinction between various subphases of Monte Albán I which they called a, b, and c. The terms "Early" and "Late" have now replaced these designations with Late I incorporating Ib and Ic. Some of the Early I forms are easily recognizable, but many of the forms now called Late Monte Albán I must also extend back into the early part of Monte Albán I.

On the other hand, the classification is too intricate in places to be effective; it is difficult to understand and more difficult to teach to people who have little or no expertise in ceramics. For this reason Blanton and Winter independently are currently working on revisions which will both simplify the system and strengthen its weak areas. Some of these analyses have been recently published (Kowalewski et al. 1978). Many suggestions made by Winter concerning vessel forms were used in sorting the collection (see below).

As mentioned, the basis for the Valley of Oaxaca ceramic typology is in the combination of the following traits: paste type, decoration, and surface finish. Clays are divided into four types: gray, yellow, cream, and brown. Originally, red, white, and black clays were recognized, but these were later seen as variants of the other four. Paddock (1962:1) notes certain attributes of these colors: the grays (gris) are mostly gray but range from white to black; the yellows (amarillos) are mostly clustered around orange and brick tones; the creams (cremas) may be quite clouded or dirty; and the browns (cafes) may be reddish brown to almost black. Various Oaxaca archeologists feel that these four basic color distinctions are fairly useless because of the difficulty in separating some of the cremas and cafes. During sorting of the Xoxocotlán material, however, less than 1% of the sherds gave immediate problems in regard to differentiating color. If one were extremely bothered by this 1%, he could always refire the sherds and match his results with what Shepard (1967:479) obtained.

Shepard (ibid.) states that the differences in color of the paste are caused by composition of the clays and methods of firing.

One type of clay is tempered with diorite and another with quartz
grains. She indicates that this may be attributed to different
centers of ceramic manufacture. This occurrence of inclusions in
various pastes aids in the sorting of sherds when color becomes a
problem. Amarillos (A) tend to become extremely fine-grained,
although there are aberrations. The cafes (K) tend to be sandy or
grainy and are usually tempered with sand rather than diorite.
Shepard (ibid.) also found that out of a sample of 577 sherds, 92%
of the crema-colored sherds (C) were tempered with diorite. It must
be mentioned that crema-ware may range in surface texture from very
rough to highly polished. Grays (G) are almost always fine-grained
and smooth and tend to be sand-tempered.

It was stated earlier that, theoretically, one characteristic of
a type is vessel form. This is partially true for the Oaxacan
classification system, although there are some forms, such as conical
cajetes, which are not limited to one phase and certainly not to one
type, as defined by paste color and surface treatment.

The manner in which La Cerámica de Monte Albán, Caso et al.'s
(1967) monumental work, is arranged makes it difficult to align types
with vessel form. Types are presented by phase; for example, under
Monte Albán I are included types G1, G2, G3, G5 . . ., C1, C2 . . .
etc. There are 39 types for Monte Albán I, some of which continue
into Monte Albán II or even later. One must then turn to another
section of the book to obtain information on vessel shapes which
occur with each type.

There are obvious shortcomings to this system, the largest one,
as mentioned, being the myriad types one must search through. The
other problem is in the basic nature of the system--without vessel
shapes being included in the actual types themselves, is the classification fulfilling its purpose? It must be admitted that there is
significance in stylistic changes, but these should be included with
characteristics such as whether the vessel is a conical, semihemispherical, cylindrical, or composite silhouette cajete, and whether it
has a rounded everted lip, a square everted lip, or no lip at all.
It should be determined whether these attributes are more chronologi-
cally diagnostic than clay composition. Winter has made advances in
this direction.

In the following pages a simple classification based on vessel
form, paste color, surface treatment, and to some extent design, is

presented for the Early Monte Albán I material. It is not inclusive of
all Early Monte Albán I material, but only includes those types found
during the work at Xoxocotlán which were considered to be diagnostic.
It was mentioned earlier that some of the vessel forms (especially the
ollas, attributed to Late Monte Albán I) must have started in Early
Monte Albán I. In fact, the few easily recognizable types, for the
most part, might be considered something other than everyday ware. It
is not possible in many cases to distinguish much of the plainer ware
from later material. It is probably the case that there is no difference in form for much of the plain ware between Early and Late Monte
Albán I.

As mentioned before, sherds are being dealt with, and not whole
vessels. Caso, Bernal, and Acosta (ibid.:16) present an interesting
fact concerning whole and broken vessels. They had both from their
excavations at Monte Albán--whole vessels from tombs, burials, and
offerings and sherds from stratigraphic tests in middens and fill.
Table 5 gives the results that they derived from plotting sherds
against whole vessels for Monte Albán I and II. One can see that gray-
ware made up the majority of ceremonial vessels in Monte Albán I and
crema-ware made up the largest percentage in Monte Albán II. This
is one example of the advantage of having whole vessels with which
to work.

Early Monte Albán I

Due to the shallowness of excavations and the frequent mixing of
Early and Late Monte Albán I deposits, only diagnostic forms are
presented. All four color types appear in Early Monte Albán I according to Caso et al., although no cafes or amarillos were recovered
which could definitely be placed in this phase.

Gray Ware
Caso, Bernal, and Acosta (1967:24-35) list Types G1, G5, G16,
G17, G18, and G24 as appearing in what they term Monte Albán Ia.
Below is a brief description of each type as it appeared in
Paddock (1962). If the description changed significantly between
1962 and 1967, when La Cerámica de Monte Albán was published, the
description is updated. Types G1, G5, and G24 are not described here
because they are not diagnostic of Early Monte Albán I exclusively.
Added to the list is Type G15, due to Winter's consideration of it

74

as a good diagnostic.

G15   Light to dark gray clay; thin to medium; polished one or both sides; may have dark gray slip or remains of red paint; decorated with plain lines or diverse motifs; incised both inside and outside.

G16   Gray clay; medium to thick; often covered with black or greenish-gray slip and generally well polished; incised decorations on exterior surface in form of cross-hatching.

G17   Gray clay; medium to thick, never thin; frequently has a black or gray-green slip; well polished; three types of decoration around border: modeled or engraved human or animal figures in bas-relief, undulating rim with no incising, and undulating rim with incising.

G18   Same as G17 except the clay is a yellowish brown, due to either a slip or to the color of the clay itself.

While these types are at times easily identifiable, they tell nothing about vessel shape. Therefore, the problem was approached from the other end. A series of descriptive categories based on vessel shape were developed first, and then surface treatment was considered as a second step.

Cajetes with 90° flaring rims (plate 15, a-g). These bowl forms always contain a right-angle rim, with accompanying decoration on the rim consisting of wavy connected lines, lines parallel to the rim, hatched lines, or punctuations. Since no complete forms were recovered, the depth of these vessels can only be approximated to be 10-12 cm. The walls of the vessels may be almost vertical to fairly outleaned. Most design elements on the rims of these vessels would fall into Types G15 and G16 of Caso, Bernal, and Acosta (1967:30-33).

Cajetes with slightly flaring rims (plates 15k and 16g, h). These vessels do not contain the 90° rims but have much less accentuated rims. They are more or less intermediate between deep cajetes with the 90° rim and the extremely outleaned cajetes described below. Decoration consists of either a double-line-break motif near the rim, with two solid lines below that followed by a wavy line, or of two solid lines near the rim separated by an area about 5 cm wide in which are placed punctuations with a stepped solid line. This latter design may follow directly from a similar one diagnostic of the Rosario phase (Winter, Drennan, personal communication). Depth of the vessels may vary from 6 to 10 cm. Many of the sherds resemble Type G16 (Caso, Bernal, and Acosta 1967:32-33).

Table 5. Percentage of color types for whole vessels and sherds from Monte Albán I and II

	GRAY	CREMA	CAFE	AMARILLO
Monte Alban I				
Sherds	25.2%	52.4%	19.8%	2.1%
Whole Vessels	76.5%	16.8%	4.3%	2.3%
Monte Alban II				
Sherds	29.1%	35.5%	15.2%	3.0%
Whole Vessels	22.2%	64.8%	8.5%	4.5%

Outleaned cajetes (plates 15h, 1; 16a-f; 20b-e, g). These cajetes have extremely outleaned walls and closely resemble soup bowls. Decoration consists of miscellaneous combinations of vertical, horizontal, and diagonal lines, similar to those of Type G16 in Caso et al. (1967).

Tecomates (plate 16i, j). The three tecomates recovered had one or more lines parallel to the rim on the exterior, followed by a wavy line or hatched lines. This motif is also similar to Type G16.

Pot stands or concentric ring vessels (plate 16n). Caso et al.(1967:191, 193) illustrate this form, which they describe as "una gran copa decorada al exterior de la base con anchas lineas grabadas paralelas; otra linea ancha se inicia en el borde interior para rematar en una gran espiral en el fondo de la copa." They also state that the clay for these vessels is Type G17. One alternate design motif consists of diamond-shaped areas free of incising surrounded by diagonal or hatched lines.

Pinched-rim cajetes (plate 16k). These vessels are outleaning forms which have had their rims pinched in three or four places, making the appearance of the vessels very lopsided. The specimens we recovered were not decorated. Caso, Bernal, and Acosta (ibid.:31) illustrate a pinched-rim vessel under Type G15.

Fish plates and eccentric rimmed vessels (plates 15i, j, m; and 17c, d). These vessels are fairly common in Early Monte Albán I deposits and are profusely illustrated in La Ceramica de Monte Albán (pp. 34, 163; plate 1), where they appear as Types G17 and G18. The rims of the plates are characterized by deep indentions and deeply incised lines running parallel to the rim. The borders between the lines may be filled with diagonal or curvilinear design motifs, sometimes simulating scales. Many times, the head of the animal being represented will appear on the rim.

Composite silhouette vessels (plates 16m; 19m; 20f). The rims on these vessels may be straight, slightly outcurved, or incurved. These forms do not have as wide an opening as the cajetes, and for this reason almost resemble tall vases. The main distinguishing characteristic is the break in body line. Decoration on one specimen consists of hatched lines, and on another, curvilinear elements. Figure 19m illustrates a modeled relief of a human on one such vessel similar to two pieces illustrated by Caso et al. (ibid.:151).

Shallow cajetes with interior differential burnishing (plate 22a-e). These vessels are of Type G5, mentioned above as occurring in Early Monte Albán I, but not described due to their occurrence in Late I as well. These G5 forms have a polished (burnished) band on the inside of the rim, the band usually being no more than 3 cm wide. The other forms which are assignable to Early Monte Albán I (Winter, personal communication) have more than just a single band: one specimen exhibits a stepped-line motif; another, a dual wavy-line motif; and a third has what appear to be trees burnished into it. These aberrant forms do not appear in La Ceramica de Monte Albán.

Shallow cajetes (?) with exterior curvilinear design and red paint (plate 19a-f). No rims were found from these vessels and, hence, they may be body sherds from another vessel form. Clay is extremely thin and vessels were probably small. Type G15 can have a red paint, as can Type G3, this latter type being described by Paddock (1962:2) as having a metallic ring when struck but no design other than the paint. The designs on the Xoxocotlán specimens are isolated floral or other curvilinear motifs.

Crema Ware

There are no crema types listed by Caso, Bernal, and Acosta (ibid.) which act as period markers for Early Monte Albán I. Winter currently has a number of crema types worked out for this early phase, which he will publish elsewhere. Therefore, only the most common marker, the Suchilquitongo bowl, is presented here.

Suchilquitongo bowls. Named by Flannery, these bowls were first recognized from excavations near the present town of Suchilquitongo, northwest of Huitzo in the Etla arm of the Valley. These vessels are shallow outleaned bowls of cream-colored (crema) clay with a garnet slip applied to a flat rim. Specimens may have a braid of applique around the outside of the rim. The bowls are supported by three nubbin feet. They are illustrated in Caso et al.(ibid.) as Type C4.

With few exceptions, the above 11 types make up the list of Early Monte Albán I diagnostics found during work near Xoxocotlán. There are more diagnostic types which did not turn up during survey and excavation. As more work is done in stratified deposits such as those which occur on residential terraces, more diagnostics will become apparent.

Late Monte Albán I

To describe all of the ceramics of this period would take up a great deal of space. Therefore, the forms discussed below represent only the more typical types recovered. Again, it must be emphasized that many of the forms described here have their origins in Early Monte Albán I and continue unchanged into Late Monte Albán I.

Gray Ware

Caso, Bernal, and Acosta (ibid.:145) list 11 gray-ware types which occur in Late Monte Albán I, not including Types G15-18. The more typical ones indicative of the phase are listed below.

G3　Dark to light gray; medium to thick. Always polished, may have remains of red powder on outside surface. Similar to Type G15.

G5　Dark to light gray in color; medium to thick. Distinguished by a polished band on the interior of the rim which averages 3 cm in width.

G13　Light gray, thin; polished over black slip. Engraved lines parallel to rim on exterior.

G25 Thin, polished gray, having a reinforcementlike flange around the base. May be a bandlike incised decoration on the body.

G26 Same as Type G25 but with the flange on the exterior near the lip. There is a variant having both base and lip moldings with a bandlike decoration between.

G30 Light gray, almost white: generally thin, well polished, sometimes over very pale slip.

There are many other types for Late Monte Albán I, but they last through Monte Albán II and many through Monte Albán III-B. Probably the most common two types are G1 and G12. Both last through Monte Albán III-B.

G1 Dark to light gray, normally thick to medium, rarely thin. From its common appearance and imperfect finish, Type G1 seems to have been in daily household use.

G12 Distinguished by two or more parallel lines on inside of rim; rarely, only one line will be present. Bottom has wavelike incisions made with a comblike instrument. Form is the very common medium-thick conical cajete. (Winter was able to show us how to distinguish Monte Albán I Type G12s from those of Monte Albán II, making this type an excellent marker for both periods.)

Again, an attempt was made to discuss vessel form:

Plain conical cajetes. Hard to distinguish from gray ware of any other phase and, consequently, a poor marker. Lip forms may be pointed or rounded, rims may be vertical or slightly flaring. A larger sample could be subdivided and the different combinations of traits grouped to determine whether they are significant. These cajetes fall under Type G1, described above.

Conical cajetes with interior rim incising and combed bottoms. This form corresponds to Type G12 described above; one difference between Monte Albán I and II forms is in the spacing of the wavelike design in the bottom of the vessel.

Small Cocijo bottles (plate 19i). These vessels, illustrated in Caso et al.(ibid.:149) are miniatures with globular bodies, flat bases, and tall necks. They usually measure between 5 and 10 cms in height and contain a well-executed relief of the face of Cocijo, the Zapotec god of rain.

Ollas with red paint under everted rim. Represented by only two specimens (plate 19k, 1). Size unknown except for mouth diameter, which is 7.5 cm on one measurable specimen.

Crema Ware

As stated earlier, for Monte Albán I as a whole, crema ware seems to have been the more common, everyday ware. This is magnified in Late I. Caso, Bernal, and Acosta (ibid.:147) list nine crema types for Monte Albán as a whole:

C1   Very pale cream; coarse texture, sandy and thick; unpolished; no decoration.

C2   Unpolished, variable thickness; distinguished from Type C1 by watery red paint, usually in bands at top and bottom and rough thick bands on the body. (Caso et al. indicate that this type does not extend into Monte Albán II.)

C3   Cream, variable thickness; distinguished from Type C1 by black smudge finish.

C4   Cream, thin to medium; well-polished garnet slip. Paddock (1962:5) and Caso et al. (ibid.:46) state that this type does not extend into Monte Albán II, although the latter authors' table VII has the type listed under Monte Albán II types.

C5   Cream; thin to medium; usually rough but at times finely polished; fine white slip which may cover only exterior and inside edge of rim; may be decorated with incised lines.

C6   Cream, thin to medium; well polished; light-yellow to reddish-brown slip.

C7   Thin to medium, polished both sides; seems to have double slip: first orange, then red; appears in upper level of Monte Albán I but more characteristic of Monte Albán II and III-A.

C13   Sandy cream; medium to thick; light-brown slip, red discoloration on upper parts. Can be confused with Type C2 but is better polished, and paint occurs in poorly distributed splotches.

C20   Sandy cream, medium to thick; well-polished shiny black slip, more common in Monte Albán II.

Of all the types listed above, only Types C2, C4, C5, and C13 are limited to Monte Albán I, according to Caso et al. (ibid.:215) and Paddock (1962:5-6). The forms listed in La Cerámica de Monte Albán as being representative of Monte Albán I fall short of including many of the forms found on the Xoxocotlán piedmont.

Ollas with varied rim styles (plates 20a; 21e, f). This group is composed of large, round vessels with restricted necks. The rims may be either vertical extensions of the neck, outcurved at a 45° angle, or rolled outward. These ollas may be plain (Type C1), painted with a splotchy red paint (Type C2), or smudged black (Type C3). They are almost never polished, and the texture of the vessels is moderately to extremely coarse. One example is of Type C7, red-on-orange polished. These ollas average about 14 to 15 cm high with an orifice diameter of 10 to 12 cm.

Tall-necked vessels. The only example recovered is a vessel of Type C1, 21 cm high with a 10-cm-wide orifice. Ten centimeters below the rim, the body wall breaks out into a globular body with a flat base.

Large open tubs. These vessels have a diameter sometimes reaching 40 to 50 cm and an estimated height of over 25 cm. Most examples are of Type C2, with two specimens of Type C1, although, as is many times the case, Type C1 sherds are really Type C2s, coming from unpainted areas of C2 vessels. This same type of tub, usually employed as a washtub, is found in the Oaxaca market today.

Large open tubs with interior handles. Identical to the vessels described above, this form has handles on the interior of the vessel to facilitate lifting.

Conical cajetes. These are fairly large-diameter vessels, ranging from 15 to 23 cm in diameter and from 7 to 13 cm in height. Lips may be pointed or round, rims may be vertical or slightly flaring. Bottoms are always flat with an interior diameter of about 12 to 13 cm.

Hemispherical cajetes with flat bases (plate 21c, d, g). These vessels are fairly large bowl forms with a height averaging 10 cm and a rim diameter of 13 to 14 cm. Forms may be of Type C6, C7, or C13, and all have the common attribute of being fairly highly polished.

Comales (plate 16 1). These are flat, platelike forms of Type C4. They always have a deeply grooved line around the interior just below the rim and three nubbin feet for supports.

Cafe Ware

Caso, Bernal, and Acosta (ibid.:49-53) list 12 types for Monte Albán I, only three of which are important here:

K1   A coarse brown clay; sandy-feeling, without polish. (This later description from Caso et al. [ibid.:49] differs radically from that of Paddock [1962:5].)

K8   Light brown to ochre; thin to medium; well polished on both sides. May have dark line painted on inside where wall joins bottom.

K19   Light brown, sandy, thin; polished on both sides over dark brown (almost black) slip.

Very few types of cafe ware were recognized, due to the paucity of cafe sherds in the area. Only two forms are discussed:

Ollas with varied neck styles. These vessels are almost identical to those made of crema clay but are a little thinner. Neck styles range from straight to flaring to rolled. Dimensions are equal to those for crema ollas. Clay used is almost exclusively Type K1, with a few examples of K19.

Conical cajetes. These vessels may have straight, out-flaring or concave, out-flaring walls. The height usually averages 8-12 cm on larger specimens and 5-6 cm on the shallower specimens. Rim diameters may vary from 14 up to 25 cm. Clays and surface treatment correspond to Types K8 and K19.

Amarillo Ware

Amarillo ware made up an insignificant percentage of the total collection; only two rim sherds were recovered. Caso et al. (ibid.:59-

60) list seven types as occurring in Monte Albán I levels. The only one listed here is Type A4.

A4   Yellow, bricklike clay; thin to medium, polished on one or both sides; often has a brick-colored slip; may have incised lines parallel to the rim on the exterior.

Only one form was distinguishable: both rim sherds were from tecomates which, due to the size of the sherds, could not be given dimensions. No vessel form was recognizable for the few body sherds recovered.

Summary

It should be obvious from the above discussion that a much greater emphasis on vessel form provides an easier-to-use classification. By sorting pottery as to form, previously existing type descriptions (which had been formulated on the basis of clay color, surface treatment, and decoration) could then be applied to the different categories to examine range of variation. In some cases, it was seen that one vessel form always corresponded to a single type as defined by Caso et al. (1967), while in other cases, one form encompassed two or more previously existing types.

Ceramic Data from Structures

Structure 14 and Feature 1

Table 6 gives the number of rim sherds by type according to Caso et al. (1967) from Area 1, Feature 1. The latter feature has been interpreted to be the midden area related to Structure 14; and due to the functional difference between the two areas, it was hypothesized that there might be a difference in ceramic types and vessel forms from the two localities. Table 7 illustrates the frequency of occurrence in each area for all of the Early Monte Albán I vessel forms, and table 8 the frequency of vessel forms from the Late Monte Albán I phase.

Table 6 shows that there were three times as many Type C1s and twice as many Type C2s in the midden as in the house. This sample of rim sherds is quite large (12 C1s and 70 C2s), but there is inherent error present due to an inability to tell how many vessels are

Table 6. Frequency of Early and Late Monte Albán I ceramic types from Structure 14 and Area 1, Feature 1 (counts include only rim sherds from subsurface deposits; surface sherds are not included)

Type	Structure 14	Area 1, Feature 1
C 1	3	9
C 2	22	48
C 3	–	1
C 4	4	4
C 5	–	16
C 6	–	10
C 7	8	1
C 13	–	1
C 20	–	1
G 1	14	25
G 5	2	–
G 12	25	12
G 15	15	8
G 16	7	1
G 17	4	6
G 18	3	1
K 1	–	2
K 8	–	1
A 4	2	–
TOTAL	109	147

Table 7. Frequency of Early Monte Albán I vessel forms from Structure 14 and Area 1, Feature 1 (counts include only rim sherds from subsurface deposits; surface sherds are not included)

	Cajetes with 90° rims	Cajetes with slightly flaring rims	Outleaned cajetes	Tecomates
Structure 14	14		1	1
Feature 1	4	2	2	

	Pot stands	Pinched-rim cajetes	Eccentric rimmed vessels	Composite silhouette vessels
Structure 14	4	2	5	1
Feature 1	5		2	1

	Shallow cajetes with interior burnishing	Shallow cajetes with exterior red paint	Suchilquitongo bowls
Structure 14	2	1	2

present.* Vessel forms corresponding to these types include all shapes of very rough-textured (<u>textura corriente</u>), rather plain-looking ollas and large tubs or wide-diameter basins. The black smudges on many of the olla sherds suggest that the vessels may have been used in cooking activities.

Three small pits were discovered chiseled into the bedrock in Area 1, Feature 1 which were certainly too small to have served as storage pits. More likely these were small hearths, suggesting that cooking was done in areas away from houses. This might help account for the large amount of olla rim sherds (49) from Area 1, Feature 1, and the smaller amount (23) from Structure 14 (table 8). While 16 square meters of Feature 1 were uncovered, as compared to only nine square meters of Structure 14, the volume of dirt removed from each was approximately equal due to the shallowness of the deposit of Feature 1. In addition, it should be remembered that while all sherds were saved from the 16 squares of Feature 1, those from six of the squares are not included in the counts because the squares were only five to eight centimeters deep. The disparity in surface areas opened up, then, is not significant, because approximately equal volumes of dirt were removed from both structures. The incongruity in numbers of olla rim sherds is even more accentuated because Feature 1 had more olla rims per volume of dirt than did Structure 14, since sherds from six of the excavation squares of Feature 1 are not included in the counts.

Since Area 1, Feature 1 has been interpreted as being a midden, it might be expected to contain more sherds than a house would, simply as a result of the casting away of broken vessels from the interior of the house. This would help explain the greater proportion of olla rims in the midden area, but does not completely answer the question of why Types C5 and C6, which are fairly fancy types, were not found in the house. Without exception, these types appeared as hemispherical cajetes with flat bases. Form actually ranged from truly hemispherical to outflaring-wall forms (not true conical forms), but there is no obvious break in the continuum from one form to another. All 26 sherds of Types C5 and C6 and, correspondingly, all 26 sherds from semihemispherical cajetes came from the midden area

---

* When a vessel was partially reconstructed, each rim sherd was counted separately.

Table 8. Frequency of Late Monte Albán I vessel forms from Structure 14 and Area 1, Feature 1 (counts include only rim sherds from subsurface deposits; surface sherds are not included)

	Vessel Form	Structure 14	Area 1, Feature 1
GRAY	Plain conical cajetes	11	21
	Conical cajetes with interior rim incising	25	12
	Unknown form	3	4
CREMA	Plain ollas with vertical neck	1	3
	Plains ollas with 45 -angle rim	9	18
	Plain ollas with rolled rim	13	28
	Tall-necked bottles	0	1
	Large, open tubs	1	4
	Large, open tubs with interior handles	0	2
	Conical cajetes	4	1
	Hemispherical cajetes with flat base	0	26
	Comales	2	4
	Unknown form	5	4
CAFE	Ollas with varied rim styles	0	2
	Conical cajetes	0	1
A	Tecomates	2	0

(tables 6 and 8). This is a rather interesting correlation. It seems odd that no hemispherical cajetes came from the house area. Even if the midden were to contain more sherds of a particular type or vessel form, it should not contain all of the sherds unless that particular vessel form was never used inside the house. Again, error might lie in an inability to determine actual numbers of vessels, but there is a fairly large sample of rims (26) spread fairly evenly between two types. It is possible that while the midden area served as a dump area, it also served as both a cooking and an eating area, the latter activity using the conical cajetes of Types C5 and C6.

The Type C1 and C2 ollas found in the house may have served as storage containers or may have been cooking vessels which had been stored in the house. Also occurring in the house is Type C7, a bichrome type for which two vessel forms can be recognized: an olla with a 45°-angle rim and a conical cajete. Eight rim sherds of this type were recovered from inside Structure 14, four making up the olla form and the other four the conical cajete. The one sherd from Feature 1 was from a conical cajete. One would be tempted to state that this distribution might be due to Type C7 vessels having been used in household ritual activity or for special occasions, but there are only three vessels present, hardly a valid sample.

The distribution of plain gray conical cajetes is identical to that of plain ollas: there are almost exactly twice as many rims in Area 1, Feature 1 as there are in Structure 14 (table 8). This again may be due to the function of Feature 1 as a midden area. On the other hand, conical cajetes which contain the double-line incised border around the interior rim (Type G12) occur in the two locales in inverse proportions, with twice as many occurring in Structure 14 as in Feature 1. In both instances there are almost an equal number of rims from both vessel forms present: 32 plain rims and 37 incised rims. The only difference lies in the inverse proportion of the numbers of each form in the two areas.

As mentioned above, no plain crema hemispherical cajetes occur in Structure 14 fill. While there are twice as many plain gray conical cajetes in Feature 1 as there are in Structure 14, 11 rims do occur in the structure (table 8). Is the distribution found due only to chance? Again, it is difficult to answer this question on the basis of sherds only and not whole vessels. If sherds are large enough, individual vessels can sometimes be distinguished by certain traits such as slight variations in rim form and projected circumference. These sherds, however, were too small to attempt this.

Caso, Bernal, and Acosta (1967:16) present percentages of whole vessels (from ceremonial contexts) and sherds (from residential debris) for Monte Albán I, for crema ware as contrasted with gray ware (table 5). They found that 76.5% of whole vessels, or, as will be inferred here, ceremonial vessels, were gray ware and 16.8% were crema ware during Monte Albán I. Crema ware, on the other hand, was much preferred for everyday usage over the gray ware. They indicate (1967:195) that the plain ware types, C1, C2, and C3, were evidently employed as cooking ware (ollas, bottles, and basins) but that other types, such as C4 and C5, which are as well finished as some of the ceremonial gray ware, were seldom recovered from tombs, offerings, or burials. They recovered only two complete vessels of Type C1, one vessel of Type C3, and only two complete ollas of Type C2 from ceremonial contexts. This contrasts with the thousands of sherds of these three types which were recovered from general fill.

Without whole vessels from ceremonial contexts, it is difficult to compare the percentages from Caso, Bernal, and Acosta with those from the Xoxocotlán piedmont structures. It is possible, however, to compare frequencies of crema-ware sherds with gray-ware sherds. Crema ware constituted 52.4% of all sherds Caso et al. found dating to the Monte Albán I phase, while gray-ware sherds made up only 25.2% of their total. Taking Structure 14 and Area 1, Feature 1 together, there are 128 crema-ware rims and 123 gray-ware rims present for Monte Albán I (Early and Late). This is quite divergent from the percentages obtained by Caso, Bernal, and Acosta, but there are biases which must be taken into account. One of these is that only rim sherds were tabulated. Crema vessels during Monte Albán I were larger than gray vessels, and the rims on the former were much smaller than the rims on the latter, due to the shape of the olla. If body sherds had been included in the totals, then the percentage of gray-ware sherds would probably have been lower in relation to crema-ware sherds.

In summary, the following can be suggested: (1) there is not an even distribution of vessel forms between the house and accompanying midden area; (2) ollas occur in greatest quantity in the midden area; this is due to the fact that it is not only a midden area but also, possibly, a cooking area; (3) plain crema cajetes occur more often in the midden (for both of the reasons stated above); (4) double-line incised cajetes occur more often in the house; (5) vessels of type C7

(a bichrome) may occur more often in the house; (6) fancy-rimmed
Early Monte Albán I vessels occur more frequently in the house area
than in the midden area.

It is not intended that the above six characteristics should be
generalized. There is only this one instance of a house and accompanying midden zone. The sample is too small to be statistically
testable. For example, table 7 lists five rims from eccentric-rimmed
vessels for Structure 14 and only two for Feature 1. The difference in
totals is probably not culturally significant.

The ceramic material from Structure 14 and its associated midden
area (Feature 1) suggests hypotheses which later excavations and
analyses can test. With a larger sample of Early and Late Monte Albán
I material from many different residences, it should be easier to
sort out patterns resulting from various distributions of vessel types.
It is obviously easier to distinguish tomb offerings when found in
that context, but in the case of ritual vessels, cooking vessels, and
eating vessels, it becomes difficult to distinguish function, due to
the facts that: (1) plain and fancier ware may both have been used for
everyday activities, and (2) one is usually dealing with small sherds.

Structure 13

A total of only 23 Late Monte Albán I and seven Early Monte Albán
I sherds were recovered from the three square meters excavated in
Structure 13. Table 9 lists ceramic types for Late Monte Albán I from
this structure. Late Monte Albán I vessel forms included ollas and
large-mouth basins in crema ware, and plain and incised conical cajetes
in gray ware (table 10). Early Monte Albán I forms included cajetes with
90° rims and eccentric-rimmed vessels.

Table 9. Frequency of Late Monte Albán I ceramic types
from Structure 13 (counts include only rim sherds
from subsurface deposits; surface sherds are not
included)

Type	Structure 13
C1	1
C2	7
C3	1
G1	8
G12	6

Table 10. Frequency of Late Monte Albán I vessel forms from
Structure 13 (counts include only rim sherds from
subsurface deposits; surface sherds are not included)

	Vessel Form	Structure 13
CREMA	Plain ollas with 45 -angle rim	4
	Plain ollas with rolled rim	4
	Large, open tubs	1
GRAY	Plain conical cajetes	8
	Conical cajetes with interior rim incising and combed bottom	6

Structure 15

Sherds from the five test pits surrounding Structure 15 were from
Early and Late Monte Albán I, Monte Albán II, and Monte Albán V (table
3). Only four Early Monte Albán I sherds and five Monte Albán V sherds
were recovered, with the rest being from the other two periods (87 Late
I and 106 II). Shortness of available excavation time and the fact
that there seemed to be mixing in all test pits deterred doing much
analysis with ceramics from Structure 15. One interesting form which
was recovered is a small olla form which probably is from Monte Albán
II (pl. 20j).

Structures 16-18

The few ceramics from these structures were not classified by
vessel form. Sherds from around Structure 16 were all from Late Monte
Albán I, from which a few semihemispherical crema cajetes were
reconstructible (Type C6). Table 4 lists the number of sherds for each
structure by phase.

FIGURINES

To a large degree, the figurines recovered from both survey and

88

excavation correspond to the descriptions published by Caso, Bernal, and Acosta (1967), and, predictably, most date to Monte Albán I, although some forms attributed to Monte Albán I last through Monte Albán II. Forms represented include humans, dogs, monkeys, and frogs. Table 11 lists each type by provenience.

Table 11. Frequency of figurine types by provenience

Figurine Type

	1a	1b	2	3	4	5	Dog	Monkey
PROVENIENCE	40a	58	15	10	29a	49a	43a	S9
	55a	S15	21a	C22	45a	41a	11b(3)	
	7	S9				C8	22a	
	29	54b				12	42a	
	F1M	3				8d	40a	
		S15				62	16	
		29b				20b	8b	
						F1M		
						8c		
						C39		
						C22		

C indicates 4-meter sample-collection circle.
F indicates feature.
M indicates midden deposit.
S indicates structure.
Other numbers are surface collection proveniences.

Type 1a

Sample: 5 (pl. 23b-c, f)
Phases: Monte Albán I and II
Form: Small human figurines, characterized by widely spread legs, short arms, and small breasts (if present). The heads are usually quite large in relation to the body, and feature upward-slanting eyes, typical of Formative figurines.

Type 1b (Caso et al. do not divide Type 1 into a and b)

Sample: 7 (pls. 22f, 23a, i)
Phases: Monte Albán I and II
Form: These pieces are larger than the Type 1a specimens but exhibit similar features. The figurines may exhibit elaborate head-pieces and highly arched eyebrows. Other facial features are variable.

Type 2

Sample: 2 (pl. 23g, h)

Phase: Monte Albán I

Form: This type is represented in the sample by only two heads. They
       are different from the head styles of the above two types,
       being very flat and wide, with eyes which are more slanted,
       either downward or upward. Both examples have thin, wide
       mouths and long, thin noses.

Type 3

Sample: 2 (pl. 22h, i)

Phase: Monte Albán I

Form: This type is also represented in the sample only by heads.
       Specimens of this type are much larger than those of Types
       1 and 2. The pieces may be highly decorated, with headdresses
       and beads, and always have more relief than specimens of Type
       2. It appears from the two specimens that the heads may
       have actually been hollow when first produced. What is left
       now is only the facial part.

Type 4

Sample: 2 (pls. 22e, 23j)

Phase: Monte Albán II

Form: Examples usually have a sneering face, upward- or downward-
       slanting eyes, molded in high relief. This type resembles
       Type 3, but the two are not easily confused. As with Type 3,
       these specimens appear to have been hollow.

Type 5

Sample: 11 (pl. 23e)

Phases: Probably Monte Albán I and II

Form: This type is similar to Type 1b but larger. Figures may be
       standing or seated. Based on comparison with excavated material
       from elsewhere in the Valley, this type probably first appears
       in the Middle Formative.

Dog Figurines

Sample: 9

Phases: According to Caso et al. these figurines occur primarily, if
         not exclusively, in Monte Albán III-A and III-B/IV.

Form: Various stylistic representations of canines.

90

Monkey Figurines

Sample: 1 (pl. 23d)

Phase: Not mentioned by Caso <u>et</u> <u>al</u>., but excavated from a Late Monte
        Albán I context.

Form: The term "monkey figurine" was decided upon for lack of a better
       term. The specimen is broken, but on the basis of the intact
       portion, it appears as if a monkey had been carved out of a
       piece of ceramic material, complete with at least two legs and
       a tail.

Based on the few pieces from survey and excavation, it is felt that
any detailed figurine classification would be presumptuous. The few
pages devoted by Caso, Bernal, and Acosta (1967) to figurine classification, even when combined with earlier work by Bernal (1947) and Caso
and Bernal (1952), has its shortcomings. This situation has been
augmented by Carl Kuttruff's (1978) work, which is, essentially, a
reclassification of all existing types. His work is based largely on
surface collections from Monte Albán, but also includes previously excavated figurines from other sites in the Valley.

8. Nonceramic Artifacts

It was decided to collect all cores and worked pieces of obsidian, chert, and flint, and either to collect or note in the field records all manos, metates, and fragments thereof. In addition, shell, whether worked or not, was collected. The reason for collecting this material was to compare it with that found on Monte Albán proper, with respect to quantity and quality. During the summer of 1973, a systematic sampling of over 100 terraces was conducted by Margie Lohse, in an effort to identify material present and also to locate any workshops which may have been present. The results of this study are not yet available. Very few obsidian blade cores have been found at Monte Albán, indicating that obsidian may have been introduced to the site in an already processed form.

CHIPPED STONE

This group consists of one projectile point plus numerous retouched blades and flakes. Table 12 lists chipped stone by collection provenience.

Projectile point

Sample: 1

Description: The specimen is 3 cm long, is made of speckled white quartzite, and has a small contracting-stem base.

Retouched or used blades (obsidian)

Sample: 8

Description: Long, parallel-edged pieces which are twice as long as they are wide, containing areas of chipping resulting from retouch or use.

Retouched or used blades (chert)

Sample: 2

Description: More or less parallel-edged pieces containing areas of chipping resulting from retouch or use.

92

Table 12. Frequency of chipped stone by provenience

	Projectile Points	Obsidian Blades	Obs. Blades End Retouch	Used Chert Blades	Used Chert Flakes
PROVENIENCE	27a	4 8c(2) C7 C8 C27 S5 F1M	C2 C22 C27 8c 10c(2)	C14 C16	C20 C29 C31 C39 11b 43a F1 F1M

	Quartzite Flakes	Sandstone Saw	Obsidian Blade Core	Flake Core	Spokeshave
PROVENIENCE	C17 C36(2) C39 2c 3 24b F1M(3)	F1	C42 24b 51a	C2	C7(2) C16(2) C39 C40(2) C43(2) 8a F1(2)

C indicates 4-meter sample-collection circle.
F indicates feature.
M indicates midden deposit.
S indicates structure.
Other numbers are surface collection proveniences.

Blades with end retouch (obsidian)

Sample: 6

Description: In addition to exhibiting lateral retouch, these pieces
exhibit steep-end retouch, forming effective scrapers.
Two of these pieces show a moderate amount of bulbar
retouch to form scraping planes.

Used flakes (chert)

Sample: 8

Description: Randomly struck pieces showing signs of slight edge
retouch. These pieces were evidently used for cutting
as opposed to scraping.

Flakes with steep retouch (quartzite)

Sample: 10

Description: These pieces exhibit steep retouch around a circular edge
adjacent to a flat surface and probably functioned as
scraper planes. These pieces are all medially struck
fragments of flake cores, and resemble what Hole, Flannery, and Neely (1969:100) refer to as core tablets.

Spokeshaves (obsidian and quartzite)

Sample: 3

Description: Two large chunks of quartzite and a piece of obsidian
exhibit side-notching, possibly as a result of being
used to smooth a wooden shaft.

Saw (sandstone)

Sample: 1

Description: The specimen is 6 cm long and has large crenulations
along one edge, resembling sawteeth.

In addition to the worked pieces described above, one obsidian
blade core, six quartzite flake cores, and five chert flake cores
were recovered.

GROUND STONE

This group consists of manos, metates, a yoke, and a digging-
stick weight. Table 13 gives the frequency of ground-stone artifacts
by provenience.

Manos

Sample: 57

Description: Usually made from basalt but may be of limestone, granite,
or diorite. Forms include square with rounded corners,
and lozenge-shaped with triangular cross-section.

Metates

Sample: 17

Description: Specimens are of either basalt or diorite. Forms include

94

flat and open (grinding slabs), and slightly trough-shaped with flat bottom.

Pestle

Sample: 1

Description: Specimen is 14 cm long and is made from an extremely soft conglomerate. The end of the handle contains a very shallow hole bored lengthwise into the shaft.

Yoke?

Sample: 1

Description: Specimen is 12 cm long (broken) and U-shaped. It is made of vesicular basalt and is similar to objects from the Gulf Coast region which have been referred to as yokes.

Digging-stick weight

Sample: 1

Description: Specimen resembles a baked-clay doughnut. It is 13 cm in outside diameter with a 4-cm hole in the middle.

Polisher

Sample: 1

Description: Small, rectangular (6 x 3 cm) limestone block which is polished on one face, possibly as a result of being used to finish the exterior of ceramic vessels.

Pendant

Sample: 1

Description: Specimen is a 2-cm-long piece of ground and semipolished obsidian in the form of a plumb bob, with a narrow incision around the neck to facilitate attachment.

Table 13. Frequency of ground stone by provenience

	Manos	Metates	Pestle	Pendant	Yoke	Weight	Polisher
	S 3	27b(3)	C 37	3b	10d	55	47	7
	S 5	29a	9					
	S 15	37b	13					
	F 1M	38a	14 (2)					
	C 7	38c	15 (2)					
	C 20	40a(3)	20a(3)					
	C 37(2)	40b	32a(2)					
	3b (2)	42b	36b					
	3c (3)	43a(2)	37a					
	10a(4)	43b	40a					
	13 (2)	47	42a					
	14 (2)	48a	S14					
	15 (3)	54 (3)						
	18	55a						
	20a(4)	65a						
	20b	66b(2)						
	21a(2)							

C indicates 4-meter sample-collection circle.
F indicates feature.
M indicates midden deposit.
S indicates structure.
Other numbers are surface collection proveniences.

WORKED SHELL

The types of shell recovered include Spondylus and Crassosstrea. Table 14 lists worked shell by provenience.

Drilled pieces

Sample: 2

Description: Both examples are tiny bits of Spondylus, each having a tiny hole through which a fastener could be passed.

Effigy pendant

Sample: 1 (pl. 22g)

Description: This specimen is a 3.5-cm-long bird-effigy pendant of Crassosstrea. Three holes were drilled in the body so that the piece could be worn about the neck.

Table 14. Frequency of shell by provenience

	Unworked shell	Drilled pieces	Effigy pendant
PROVENIENCE	20b	Cut 4	F 1M
	62	7	
	58		
	41b		
	C16		
	Cut 4		

F indicates feature.
M indicates midden.
C indicates 4-meter sample-collection circle.
Other numbers are surface collection proveniences.

SUMMARY

Nothing conclusive was derived from studying the distribution of the nonceramic artifact types, except for the observation that the frequencies for chipped stone were higher in the 4 m collection circles. This would be expected since these were more intensively collected. Area 1, Feature 1 contained a high percentage of utilized pieces, which would be expected in a midden area. The distribution of manos and metates is definitely inconclusive. There seems to be little correlation with areas of high ceramic density.

9. Burials

The remains of seven individuals were recovered, but of these, only three were primary burials. Two of the seven individuals are represented by a single bone each. One of these is an adult toe phalange found in Area 1, Feature 1 (square 1N1E), and the other is a juvenile left radius found in the midden deposit associated with Structure 16. These two were not given burial numbers.

Burial 1

Location: On the south side of the lower tier of Structure 9 (pl. 7).

Orientation: Extended, 8° east of north; supine position, facing up, arms alongside body.

Age and Sex: 17-20 ± 2 years, male.

Features: Small tomb constructed over upper half of body.

Preservation: Poor, many bones crushed; lower torso, legs, and hands missing.

Discussion: As mentioned in chapter VI, this individual had been placed at right angles to the tier, so that only the upper half of the body was covered by a small tomb. As a result, all bones from the lumbar vertebrae down were missing, as were all wrist bones and finger bones. The radii, ulnae, humeri, clavicles, left scapula, and a few thoracic and cervical vertebrae remained, along with the cranium and mandible (pl. 8).

Lack of tooth wear and the fact that the third molars were just emerging argue for an age of between 17 and 20 ± 2 years. Also, suture closure (a highly variable factor) and the fact that the distal epiphysis of the right radius had not united lend support to this age estimate. The bones are not very robust, although the humeri are quite long for the Valley of Oaxaca series (Dick Wilkinson, personal communication). The skull is thin-walled and the mastoids are small, but the mandible is well within the male morphology.

Although stature reconstruction using arm bones is risky

(Bass 1971:125; Trotter and Glesser 1958:120), the right radius is the only complete bone available from the individual. Trotter and Glesser (1952) give a formula for reconstructing the stature of males using the radius, and with the application of this index the projected stature is between 59 and 63 inches.

Burial 2

Location: In central area of Structure 14.

Orientation: Unknown, very few bones present.

Age and Sex: Midtwenties, male.

Features: None

Preservation: Poor, most of skeleton missing.

Discussion: The individual is represented by the right half of a mandible and a few postcranial fragments, including a few fragments of both humeri. Both humeri exhibit large deltoid crests. Tooth wear is not severe, but several teeth indicate incomplete enamel formation.

Burial 3

Location: Adjacent to Burial 2 in central area of Structure 14.

Orientation: Unknown, only one bone present.

Age and Sex: Subadult (20 ± 2 years), female.

Features: None

Preservation: Poor, most of skeleton missing.

Discussion: Only the left side of the mandible remains, but slight tooth wear indicates an age approaching 20. Morphology is in the female range.

Burial 4

Location: Midden deposit outside Structure 15 (not connected with Structure 15).

Orientation: Unknown, very few bones present.

Age and Sex: Old adult, male.

Features: None

Preservation: The two bones present are in excellent condition.

Discussion: The mandible and left tibia are all that remain from what appears to be a secondary burial. Bone had completely fused over all tooth cavities on the mandible, except for the canines, a clue to the advanced age of the individual. The tibia is grossly pathological, character- ized by anterior bowing, swelling of the anterior edge and proximal quarter, and longitudinal striations and porosis of the outer table of the bone. The disease worsened closer to the knee.

Burial 5

Location: Twenty centimeters below Burial 4 in midden deposit outside Structure 15.

Orientation: Unknown, very few bones present.

Age and Sex: Young adult, male

Features: None.

Preservation: Poor, very few bones present.

Discussion: The mandible, fragments of a frontal, occipital, and
temporal bone, and a left radius are all that remain
from a secondary burial. None of the bones are pathological.

10. Statistical Summary of Surface Collections

This chapter is designed to provide a summary statement concerning statistical techniques and data obtained during the course of the Xoxocotlán Project. As noted previously, two types of collections were made: "survey collections" (chapter II) and "sample-circle collections" (chapter VI). The following discussion of the statistical operations and resulting data is based on distinctions made between these two types of collections.

Analysis of the "survey collection" portion of the project was undertaken in an attempt to discern patterns in occupational density through time within the region. Tables 15 through 18 represent a descriptive summary of the ceramic material collected during the survey-collection portion of the project. Each table is largely self-explanatory. The coding of the variables included in the computer operations is presented in table 19. Each of the 29 variables is identified as to format, column space, and the name or description of the variable. The Spearman Rank-order Test (Nie, Bent, and Hull 1970:153-156) was used to produce correlation coefficients between the variables listed in Table 19. These inferential statistics can be used to locate possible biases in the collecting procedures of the survey (see chapter XI) as well as to relate the distribution of various archaeological classes to one another. The Spearman Rank-order Test is a nonparametric, inferential statistic. Because of the few assumptions of this test in regard to the normalized distribution of the population being examined, it appears to be the most appropriate test available for the analysis of the Xoxocotlán material.

At this point, it should be noted that all of the ceramic material was analyzed using the system of computer programs entitled "Statistical Package for the Social Sciences" (SPSS). Calculation of density values,

Table 15. General summary of ceramic survey collections

Collection Area	Collection Unit	Size M2	Vegetation Type	Total number of sherds	Total sherd density/Ha
1	1	9950	1	82	82
2	2	6325	3	128	202
3a	3	3175	2	22	69
3b	4	4150	2	18	43
3c	5	3575	2	16	45
3d	6	4225	2	18	43
4a	7	7050	6	59	84
4b	8	2875	6	0	0
5a	9	4100	1	74	180
5b	10	2550	1	75	294
6	11	5225	1	0	0
7	12	11000	7	19	17
8a	13	1650	1	10	61
8b	14	1550	1	2	13
8c	15	2900	1	49	169
8d	16	3475	1	69	199
8e	17	1825	1	18	99
9	18	975	1	21	215
10a	19	2900	3	15	52
10b	20	2250	3	31	138
10c	21	1550	3	21	135
11a	22	3575	4	31	87
11b	23	4525	4	87	192
12	24	4225	4	32	76
13	25	1350	2	33	244
14	26	2825	4	14	50
15	27	1150	2	39	339
16	28	2275	2	19	84
17	29	3300	4	9	27
18	30	2900	2	14	48
19a	31	1600	7	3	19
19b	32	2000	5	7	35
19c	33	1825	6	29	159
20a	34	3750	6	32	85
20b	35	3950	6	25	63
21a	36	2925	1	60	205
21b	37	3025	1	55	182
22a	38	1900	7	6	32
22b	39	3325	7	4	17
23	40	2775	7	0	0
24a	41	4800	7	25	52
24b	42	4600	7	48	104
25a	43	4400	7	15	34
25b	44	4700	7	19	40
26	45	9550	7	18	19
27a	46	6825	6	20	29
27b	47	5500	6	28	51
28a	48	1625	2	13	80
28b	49	1625	5	0	0
29a	50	5725	6	40	70
29b	51	4725	6	87	184
30	52	4250	2	17	40
31	53	4425	3	20	45
32a	54	10750	6	6	6
32b	55	11025	6	9	8
33a	56	2025	6	14	69
33b	57	1325	6	0	0
34	58	6550	6	4	6
35	59	9125	4	16	18
36a	60	7975	4	58	73
36b	61	7350	4	18	24

Table 15. General summary of ceramic survey collections (cont.)

Collection Area	Collection Unit	Size M2	Vegetation Type	Total number of sherds	Total sherd density/Ha
37a	62	4775	3	178	373
37b	63	10075	6	95	94
37c	64	7525	6	0	0
38a	65	5625	2	24	43
38b	66	4000	2	81	202
38c	67	3425	2	31	91
39a	68	1775	3	140	789
39b	69	2050	3	31	151
40a	70	3150	3	343	1089
40b	71	3450	3	64	186
41a	72	2800	3	117	413
41b	73	3400	3	34	100
42a	74	4975	2	180	362
42b	75	5050	2	108	214
43a	76	5900	6	163	276
43b	77	4400	6	42	95
44a	78	5675	2	32	56
44b	79	4100	2	78	190
45a	80	3575	2	19	53
45b	81	2225	2	24	108
46a	82	1275	4	60	471
46b	83	1300	3	15	115
47	84	5950	3	172	289
48a	85	3175	5	18	57
48b	86	2000	5	10	50
49a	87	3350	3	7	21
49b	88	3225	3	10	31
50	89	8050	2	17	21
51a	90	1650	3	63	382
51b	91	1525	3	32	210
51c	92	2925	3	65	222
52	93	4625	6	42	91
53	94	1775	2	95	535
54a	95	2725	2	108	396
54b	96	3850	2	16	42
55a	97	1975	3	39	197
55b	98	4425	3	61	138
56	99	4325	6	24	55
57a	100	3575	2	16	45
57b	101	1200	2	7	58
58	102	5150	2	27	52
59	103	1175	2	4	34
60	104	2900	2	11	38
61	105	16825	5	0	0
62	106	3125	2	21	67
63	107	2775	2	8	29
64a	108	2400	6	2	8
64b	109	2275	5	5	22
64c	110	2300	6	0	0
65a	111	4700	3	25	53
65b	112	4150	3	10	24
66a	113	3550	5	7	20
66b	114	3900	5	6	15
67	115	8275	6	8	10
68a	116	3200	2	24	75
68b	117	3575	2	11	31
69a	118	3425	6	7	21
69b	119	3200	6	6	19
70a	120	2325	2	6	26
70b	121	2325	2	6	26

Table 16. Number of sherds per survey-collection unit by phase

Collection Area	Collection Unit	Early I	Late I	II	III-A	III-B/IV	V	Unknown
1	1	12	43	6	0	0	2	19
2	2	9	72	8	3	0	2	34
3a	3	5	16	1	1	0	0	1
3b	4	3	15	0	0	0	0	0
3c	5	3	13	0	0	0	0	0
3d	6	4	14	0	0	0	0	0
4a	7	1	28	7	0	0	0	23
4b	8	0	0	0	0	0	0	0
5a	9	5	66	3	0	0	0	0
5b	10	4	46	4	2	0	0	19
6	11	0	0	0	0	0	0	0
7	12	5	14	0	0	0	0	0
8a	13	2	4	0	0	0	0	4
8b	14	0	2	0	0	0	0	0
8c	15	9	17	0	4	1	5	13
8d	16	8	32	1	4	0	0	24
8e	17	4	5	0	6	1	0	2
9	18	3	9	0	1	1	3	4
10a	19	5	4	0	1	0	0	5
10b	20	5	11	3	0	3	1	8
10c	21	2	8	2	0	0	0	9
11a	22	3	13	2	3	0	0	10
11b	23	10	35	5	3	4	8	22
12	24	3	12	2	1	0	3	11
13	25	1	18	0	0	0	5	9
14	26	0	3	0	0	2	1	8
15	27	0	16	3	2	4	3	11
16	28	0	6	2	1	1	2	7
17	29	0	3	2	1	0	0	3
18	30	0	3	1	0	0	3	7
19a	31	0	0	0	0	1	0	2
19b	32	1	0	0	0	0	1	5
19c	33	0	13	5	0	0	0	11
20a	34	3	16	3	0	0	1	9
20b	35	0	13	0	0	2	0	10
21a	36	6	37	4	0	0	0	13
21b	37	5	30	3	0	1	0	16
22a	38	0	3	1	0	1	1	0
22b	39	0	3	0	0	0	0	1
23	40	0	0	0	0	0	0	0
24a	41	0	7	2	1	6	2	7
24b	42	2	24	2	1	2	3	14
25a	43	3	4	0	0	0	0	8
25b	44	1	2	2	1	3	0	10
26	45	1	4	1	0	4	0	8
27a	46	0	10	2	2	0	0	6
27b	47	1	7	3	3	3	0	11
28a	48	0	9	1	0	0	0	3
28b	49	0	0	0	0	0	0	0
29a	50	4	20	0	0	0	1	15
29b	51	8	43	6	0	0	4	26
30	52	0	6	1	0	1	3	6
31	53	2	7	0	0	0	3	8
32a	54	0	3	1	0	0	0	2
32b	55	0	2	0	0	0	2	5
33a	56	0	5	0	0	0	5	4
33b	57	0	0	0	0	0	0	0
34	58	0	4	0	0	0	0	0
35	59	0	8	0	0	0	0	8
36b	61	0	16	1	0	0	1	0
37a	62	9	105	11	0	0	1	52

Table 16. Number of sherds per survey-collection unit by phase (cont.)

Collection Area	Collection Unit	Early I	Late I	II	III-A	III-B/IV	V	Unknown
37b	63	8	51	9	0	0	6	21
37c	64	0	0	0	0	0	0	0
38a	65	4	10	1	0	1	0	8
38b	66	6	41	6	0	3	0	25
38c	67	5	16	1	0	0	0	9
39a	68	7	64	5	6	8	0	50
39b	69	0	11	3	0	1	0	16
40a	70	32	212	6	0	11	0	82
40b	71	3	35	0	0	0	0	26
41a	72	12	74	2	0	0	1	28
41b	73	1	23	0	0	0	1	9
42a	74	7	96	19	0	1	0	57
42b	75	6	65	1	1	5	0	30
43a	76	9	102	11	2	0	7	32
43b	77	4	17	3	0	0	5	13
44a	78	2	12	1	0	0	8	9
44b	79	2	35	6	1	0	13	21
45a	80	0	7	2	0	0	4	6
45b	81	0	8	3	0	2	1	12
46a	82	2	44	7	0	0	7	0
46b	83	0	7	1	0	0	2	5
47	84	7	98	15	0	0	10	42
48a	85	0	9	2	0	0	2	5
48b	86	0	1	0	0	0	4	5
49a	87	0	2	2	0	0	2	1
49b	88	0	4	1	0	0	1	4
50	89	0	2	0	0	0	11	4
51a	90	4	32	3	0	0	2	22
51b	91	3	16	3	0	0	5	5
51c	92	4	32	6	0	1	5	18
52	93	5	20	6	0	2	1	9
53	94	3	45	7	0	0	15	25
54a	95	7	64	14	0	0	2	21
54b	96	1	11	2	0	0	0	2
55a	97	1	25	5	0	1	0	7
55b	98	5	38	0	0	0	3	15
56	99	1	6	3	0	9	0	5
57a	100	0	8	1	0	7	0	0
57b	101	0	2	1	0	0	2	2
58	102	2	8	3	0	7	0	7
59	103	0	0	0	0	0	1	3
60	104	0	4	0	0	0	5	2
61	105	0	0	0	0	0	0	0
62	106	0	4	1	0	0	9	7
63	107	0	1	0	0	0	4	3
64a	108	0	0	1	0	0	1	0
64b	109	0	0	0	0	0	3	2
64c	110	0	0	0	0	0	0	0
65a	111	1	15	0	0	0	1	8
65b	112	0	3	0	0	0	3	4
66a	113	0	2	0	0	0	0	5
66b	114	0	4	0	0	0	0	2
67	115	0	1	0	0	0	2	5
68a	116	2	11	0	0	2	0	9
68b	117	1	5	1	0	0	0	4
69a	118	0	1	1	0	0	0	5
69b	119	0	0	1	0	0	0	5
70a	120	0	0	0	0	0	4	2
70b	121	0	1	0	0	0	2	2

Table 17. Ceramic density (rim sherds per hectare) of survey-collection units by phase

Collection Area	Collection Unit	Early I	Late I	II	III-A	III-B/IV	V	Unknown
1	1	12	43	6	0	0	2	19
2	2	14	114	13	5	0	3	54
3a	3	16	50	3	3	0	0	3
3b	4	7	36	0	0	0	0	0
3c	5	8	36	0	0	0	0	0
3d	6	9	33	0	0	0	0	0
4a	7	1	40	10	0	0	0	33
4b	8	0	0	0	0	0	0	0
5a	9	12	160	7	0	0	0	0
5b	10	16	180	16	8	0	0	75
6	11	0	0	0	0	0	0	0
7	12	5	13	0	0	0	0	28
8a	13	12	24	0	0	0	0	24
8b	14	0	13	0	0	0	0	0
8c	15	31	59	0	14	3	17	45
8d	16	23	92	3	12	0	0	69
8e	17	22	27	0	33	5	0	11
9	18	31	92	0	10	10	31	41
10a	19	17	14	0	3	0	0	17
10b	20	22	49	13	0	13	4	36
10c	21	13	52	13	0	0	0	58
11a	22	8	36	6	8	0	0	28
11b	23	22	77	11	7	9	18	49
12	24	7	28	5	2	0	7	26
13	25	7	133	0	0	0	37	67
14	26	0	11	0	0	7	4	28
15	27	0	139	26	17	35	26	96
16	28	0	26	9	4	4	9	31
17	29	0	9	6	3	0	0	9
18	30	0	10	3	0	0	10	24
19a	31	0	0	0	0	6	0	12
19b	32	5	0	0	0	0	5	25
19c	33	0	71	27	0	0	0	60
20a	34	8	43	8	0	0	3	24
20b	35	0	33	0	0	5	0	25
21a	36	21	126	14	0	0	0	44
21b	37	17	99	10	0	3	0	53
22a	38	0	16	5	0	5	5	0
22b	39	0	13	0	0	0	0	4
23	40	0	0	0	0	0	0	0
24a	41	0	15	4	2	12	4	15
24b	42	4	52	4	2	4	7	30
25a	43	7	9	0	0	0	0	18
25b	44	2	4	4	2	6	0	21
26	45	1	4	1	0	4	0	8
27a	46	0	15	3	3	0	0	9
27b	47	2	13	5	5	5	0	20
28a	48	0	55	6	0	0	0	18
28b	49	0	0	0	0	0	0	0
29a	50	7	35	0	0	0	2	26
29b	51	17	91	13	0	0	8	55
30	52	0	14	2	0	2	7	14
31	53	5	16	0	0	0	7	18
32a	54	0	3	1	0	0	0	2
32b	55	0	2	0	0	0	2	5
33a	56	0	25	0	0	0	25	20
33b	57	0	0	0	0	0	0	0
34	58	0	6	0	0	0	0	0
35	59	0	9	0	0	0	0	9
36a	60	6	39	3	0	0	3	23
36b	61	0	22	1	0	1	0	0

Table 17. Ceramic density (rim sherds per hectare) of survey-collection units by phase (cont.)

Collection Area	Collection Unit	Early I	Late I	II	III-A	III-B/IV	V	Unknown
37a	62	19	220	23	0	0	2	109
37b	63	8	51	9	0	0	6	21
37c	64	0	0	0	0	0	0	0
38a	65	7	18	2	0	2	0	14
38b	66	15	102	15	0	7	0	62
38c	67	15	47	3	0	0	0	26
39a	68	39	361	28	34	45	0	282
39b	69	0	54	15	0	5	0	78
40a	70	102	673	19	0	35	0	260
40b	71	9	101	0	0	0	0	75
41a	72	43	264	7	0	0	4	100
41b	73	3	68	0	0	0	3	26
42a	74	14	193	38	0	2	0	115
42b	75	12	129	2	2	10	0	59
43a	76	15	173	19	3	0	12	54
43b	77	9	39	7	0	0	11	30
44a	78	4	21	2	0	0	14	16
44b	79	5	85	15	2	0	32	51
45a	80	0	20	6	0	0	11	17
45b	81	0	36	13	0	0	4	54
46a	82	16	345	55	0	0	55	0
46b	83	0	54	8	0	0	15	38
47	84	12	165	25	0	0	17	71
48a	85	0	28	6	0	0	6	16
48b	86	0	5	0	0	0	20	25
49a	87	0	6	6	0	0	6	3
49b	88	0	12	3	0	0	3	12
50	89	0	2	0	0	0	14	5
51a	90	24	194	18	0	0	12	133
51b	91	20	105	20	0	0	33	33
51c	92	14	109	21	0	5	17	62
52	93	11	43	13	0	0	4	19
53	94	17	254	39	0	0	85	141
54a	95	26	235	51	0	0	7	77
54b	96	3	29	5	0	0	0	5
55a	97	5	127	25	0	5	0	35
55b	98	11	86	0	0	0	7	34
56	99	2	14	7	0	21	0	12
57a	100	0	22	3	0	20	0	0
57b	101	0	17	8	0	0	17	17
58	102	4	16	6	0	14	0	14
59	103	0	0	0	0	0	9	26
60	104	0	14	0	0	0	17	7
61	105	0	0	0	0	0	0	0
62	106	0	13	3	0	0	29	22
63	107	0	4	0	0	0	14	11
64a	108	0	0	4	0	0	4	0
64b	109	0	0	0	0	0	13	9
64c	110	0	0	0	0	0	0	0
65a	111	2	32	0	0	0	2	17
65b	112	0	7	0	0	0	7	10
66a	113	0	6	0	0	0	0	14
66b	114	0	10	0	0	0	0	5
67	115	0	1	0	0	0	2	6
68a	116	6	34	0	0	6	0	28
68b	117	3	14	3	0	0	0	11
69a	118	0	3	3	0	0	0	15
69b	119	0	0	3	0	0	0	16
70a	120	0	0	0	0	0	17	9
70b	121	0	4	0	0	0	9	9

Table 18. Percent of ceramics representing each phase in survey collections

Collection Area	Collection Unit	Early I	Late I	II	IIIA	IIIB/IV	V	Unknown
1	1	14	52	7	6	0	2	23
2	2	7	56	6	2	0	2	27
3a	3	23	73	5	5	0	0	5
3b	4	17	83	0	0	0	0	0
3c	5	19	81	0	0	0	0	0
3d	6	22	78	0	0	0	0	0
4a	7	2	47	12	0	0	0	39
4b	8	0	0	0	0	0	0	0
5a	9	6	89	9	0	0	0	0
5b	10	5	51	5	3	0	0	25
6	11	0	0	0	0	0	0	0
7	12	82	0	0	0	0	0	0
8a	13	20	40	0	0	0	0	40
8b	14	0	100	0	0	0	0	0
8c	15	18	35	0	8	2	10	27
8d	16	12	46	1	6	0	0	35
8e	17	22	28	0	33	6	0	11
9	18	14	43	0	5	5	14	19
10a	19	33	27	0	7	0	0	33
10b	20	16	35	10	0	10	3	26
10c	21	10	38	10	0	0	0	43
11a	22	10	42	6	10	0	0	32
11b	23	11	40	6	3	5	9	25
12	24	9	38	6	3	0	9	34
13	25	3	55	0	0	0	15	27
14	26	0	21	0	0	14	7	57
15	27	0	41	8	5	10	8	28
16	28	0	32	11	5	5	11	37
17	29	0	33	22	11	0	0	33
18	30	0	21	7	0	0	21	50
19a	31	0	0	0	0	33	0	67
19b	32	14	0	0	0	0	14	71
19c	33	0	45	17	0	0	0	38
20a	34	9	50	9	0	0	3	28
20b	35	0	52	0	0	8	0	40
21a	36	10	62	7	0	0	0	22
21b	37	9	55	5	0	2	0	29
22a	38	0	50	17	0	17	17	0
22b	39	0	75	0	0	0	0	25
23	40	0	0	0	0	0	0	0
24a	41	0	28	8	4	24	8	28
24b	42	4	50	4	2	4	6	29
25a	43	20	27	0	0	0	0	53
25b	44	5	11	11	5	16	0	53
26	45	6	22	6	0	22	0	44
27a	46	0	50	10	10	0	0	30
27b	47	4	25	11	11	11	0	39
28a	48	0	69	8	0	0	0	23
28b	49	0	0	0	0	0	0	0
29a	50	10	50	0	0	0	2	38
29b	51	9	49	7	0	0	5	30
30	52	0	35	6	0	6	18	35
31	53	10	35	0	0	0	15	40
32a	54	0	50	17	0	0	0	33
32b	55	0	22	0	0	0	22	56
33a	56	0	36	0	0	0	36	29
33b	57	0	0	0	0	0	0	0
34	58	0	100	0	0	0	0	0
35	59	0	50	0	0	0	0	50
36a	60	9	53	3	0	0	3	31
36b	61	0	88	6	0	0	6	0

Page 109.

Table 18. Percent of ceramics representing each phase in survey collections (cont.)

Collection Area	Collection Unit	Early I	Late I	II	IIIA	IIIB/IV	V	Unknown
37a	62	5	59	6	0	0	1	29
37b	63	8	54	9	0	0	6	22
37c	64	0	0	0	0	0	0	0
38a	65	17	42	4	0	4	0	33
38b	66	7	51	7	0	4	0	31
38c	67	16	52	3	0	0	0	29
39a	68	5	46	4	4	6	0	36
39b	69	0	35	10	0	3	0	52
40a	70	9	62	2	0	3	0	24
40b	71	5	55	0	0	0	0	41
41a	72	10	63	2	0	0	1	24
41b	73	3	68	0	0	0	3	26
42a	74	4	53	11	0	1	0	32
42b	75	6	60	1	1	5	0	28
43a	76	6	63	7	1	0	4	20
43b	77	10	40	7	0	0	12	31
44a	78	6	38	3	0	0	25	28
44b	79	3	45	8	1	0	17	27
45a	80	0	37	11	0	0	21	32
45b	81	0	33	12	0	3	4	50
46a	82	3	73	12	0	0	12	0
46b	83	0	47	7	0	0	13	33
47	84	4	57	9	0	0	6	24
48a	85	0	50	11	0	0	11	28
48b	86	0	10	0	0	0	40	50
49a	87	0	29	29	0	0	29	14
49b	88	0	40	10	0	0	10	40
50	89	0	12	0	0	0	65	24
51a	90	6	51	5	0	0	3	35
51b	91	9	50	9	0	0	16	16
51c	92	6	49	9	0	1	8	28
52	93	12	48	14	0	5	3	21
53	94	3	47	7	0	0	16	26
54a	95	6	59	13	0	0	2	19
54b	96	6	69	13	0	0	0	13
55a	97	3	64	13	0	3	0	18
55b	98	8	62	0	0	0	5	25
56	99	4	25	13	0	38	0	21
57a	100	0	50	6	0	44	0	0
57b	101	0	29	14	0	0	29	29
58	102	7	30	11	0	26	0	26
59	103	0	0	0	0	0	25	75
60	104	0	36	0	0	0	45	18
61	105	0	0	0	0	0	0	0
62	106	0	19	5	0	0	43	33
63	107	0	13	0	0	0	50	38
64a	108	0	0	50	0	0	50	0
64b	109	0	0	0	0	0	60	40
64c	110	0	0	0	0	0	0	0
65a	111	4	60	0	0	0	4	32
65b	112	0	30	0	0	0	30	46
66a	113	0	29	0	0	0	0	71
66b	114	0	67	0	0	0	0	33
67	115	0	13	0	0	0	25	63
68a	116	8	46	0	0	8	0	38
68b	117	9	45	9	0	0	0	36
69a	118	0	15	15	0	0	0	70
69b	119	0	0	17	0	0	0	83
70a	120	0	0	0	0	0	67	33
70b	121	0	17	0	0	0	33	33

Table 19. Coding of variables used in statistical analysis of the survey collections

Variable	Name of Variable
1	Collection area
2	Subarea designation
3	Collection unit number
4	Size of unit in square meters
5	Type of vegetation on unit
6	Estimated percent of ground cover
7	Total number of sherds
8	Number of Early MA I sherds
9	Number of Late MA I sherds
10	Number of MA II sherds
11	Number of MA III-A sherds
12	Number of MA III-B/IV sherds
13	Number of MA V sherds
14	Number of sherds not identified by phase
15	Total sherd density of collection unit
16	Density of Early MA I phase
17	Density of Late MA I phase
18	Density of MA II phase
19	Density of MA III-A phase
20	Density of MA III-B/IV phase
21	Density of MA V phase
22	Density of sherds not identified by phase
23	Percent of Early MA I phase
24	Percent of Late MA I phase
25	Percent of MA II phase
26	Percent of MA III-A phase
27	Percent of MA III-B/IV phase
28	Percent of MA V phase
29	Percent of sherds not identified by phase

correlation coefficients, and percentages was performed by using the
SPSS programs. Full details on the operating procedures and restric-
tions of these programs are available in Nie, Bent, and Hull (1970).

Computer maps discussed in a later chapter were produced by a
program written by the Laboratory for Computer Graphics, Harvard
University (Schmidt 1973). This program, SYMAP, is on file at the
Computer Center, University of Washington, Seattle, Washington. The
version used in this report has been modified by the Department of
Geography, University of Washington. Adaptation of the program for
analysis of archaeological material was undertaken by Jerry V. Jermann,
Department of Anthropology, University of Washington. The presentation
of the computer maps and analysis of the distributional patterns
appear in chapter XII.

During the course of ceramic analysis, accuracy of the laboratory
techniques was assured by recounting and reanalyzing sherds. Sherd
counts were checked several times on computer printouts to make sure
there had been no errors in transcribing counts from paper to the
80-column computer punch cards. The area of each collection unit was
calculated twice by the same individual. Collection units were traced
directly from the aerial photograph onto millimeter-square graph paper.
The area of each collection unit was carefully calculated by means of
this gridded map system. The fact that the same person calculated
all the areas assures that procedures were standardized and uniform.

The descriptive summary of the artifacts collected from the
four-meter circles during the sampling of the areas prior to excavation, comprises the second set of material discussed below. Table
20 presents the total number of sherds for each sample collection, the
density value of the total number of sherds, and the number of sherds
representing each cultural phase. Table 21 summarizes the density
values of each sample collection according to the cultural phase.
Table 22 presents the percentage of sherds representing each phase
present within the four-meter-circle sample collections.

Throughout analysis of the collections, vegetational characteristics present on the surface at the time of collection were considered.
The following coding system was employed in ranking the types of
vegetation:

Vegetation Code	Field Description of Type of Vegetation on Surface
1.	Plowed field, little or no vegetation on surface

Table 20. Summary of total number of sherds, total density, and number of sherds by phase of 4-meter sample-circle collections

Sample Coll	Total Sherds	Total Density	Early I	Late I	II	III-A	III-B/IV	V	Unknown
1	11	8751	1	10	0	0	0	0	0
2	42	33413	2	26	4	0	0	1	9
3	15	11933	0	14	1	0	0	0	0
4	28	22275	1	12	1	0	4	5	5
5	97	77168	7	85	0	0	0	0	5
6	45	35800	5	35	2	0	0	1	2
7	37	29435	4	29	1	0	1	0	2
8	37	29435	3	24	1	0	0	1	8
9	56	44551	0	52	0	0	0	3	1
10	30	23866	1	28	0	0	0	1	0
11	28	22275	2	22	0	0	0	0	4
12	19	15115	4	15	0	0	0	0	0
13	46	36595	2	33	2	0	1	1	7
14	29	23071	1	23	0	0	0	1	4
15	12	9547	0	9	1	0	0	0	2
16	24	19093	1	18	1	0	0	1	3
17	35	27844	1	27	3	0	0	0	4
18	NUMBER			NOT		UTILIZED			
19	NUMBER			NOT		UTILIZED			
20	NUMBER			NOT		UTILIZED			
21	6	4773	0	5	1	0	0	0	0
22	53	42164	0	34	3	0	7	0	9
23	0	0	0	0	0	0	0	0	0
24	2	1591	0	0	0	0	0	0	2
25	NUMBER			NOT		UTILIZED			
26	5	3978	0	5	0	0	0	0	0
27	14	11138	2	10	0	0	0	0	2
28	25	19889	0	18	0	0	0	0	7
29	30	23866	3	23	0	0	0	0	4
30	27	21480	1	20	0	0	1	0	5
31	62	49324	3	42	0	0	1	0	16
32	0	0	0	0	0	0	0	0	0
33	16	12729	3	13	0	0	0	0	0
34	2	1591	0	2	0	0	0	0	0
35	42	33413	0	30	1	0	2	0	9
36	62	49324	0	44	6	0	3	0	9
37	80	63644	5	65	1	0	1	0	8
38	97	77168	5	86	2	0	0	0	4
39	11	8751	0	7	2	0	0	0	2
40	19	15115	0	15	0	0	0	2	2
41	12	9547	0	9	0	0	0	3	0
42	42	33413	1	38	0	0	1	0	2
43	21	16706	2	19	0	0	0	0	0
44	19	15115	3	15	0	0	0	0	1

Table 21. Density values of sherds by phase for 4-meter sample-circle collections

Sample	Early I	Late I	II	III-A	III-B/IV	V	Unknown
1	796	7955	0	0	0	0	0
2	1591	20684	3182	0	0	796	7160
3	0	11138	796	0	0	0	0
4	796	9547	796	0	3182	3978	3978
5	5569	67621	0	0	0	0	3978
6	3978	27844	1591	0	0	796	1591
7	3182	23071	796	0	796	0	1591
8	2387	19093	796	0	0	796	6364
9	0	41368	0	0	0	2387	796
10	796	22275	0	0	0	796	0
11	1591	17502	0	0	0	0	3182
12	3182	11933	0	0	0	0	0
13	1591	26253	1591	0	796	796	5569
14	796	18298	0	0	0	796	3182
15	0	7160	796	0	0	0	1591
16	796	14320	796	0	0	796	2387
17	796	21480	2387	0	0	0	3182
18	NUMBER		NOT		UTILIZED		
19	NUMBER		NOT		UTILIZED		
20	NUMBER		NOT		UTILIZED		
21	0	3978	796	0	0	0	0
22	0	27049	2387	0	5569	0	7160
23	0	0	0	0	0	0	0
24	0	0	0	0	0	0	1591
25	0	3978	0	0	0	0	0
26	NUMBER		NOT		UTILIZED		
27	1591	7955	0	0	0	0	1591
28	0	14320	0	0	0	0	5569
29	2387	18298	0	0	0	0	3182
30	796	15911	0	0	796	0	3978
31	2387	33413	0	0	796	0	12729
32	0	0	0	0	0	0	0
33	2387	10342	0	0	0	0	0
34	0	1591	0	0	0	0	0
35	0	23866	796	0	1591	0	7160
36	0	35004	4773	0	2387	0	7160
37	3978	51710	796	0	796	0	6364
38	3978	68417	1591	0	0	0	3182
39	0	5569	1591	0	0	0	1591
40	0	11933	0	0	0	1591	1591
41	0	7160	0	0	0	2387	0
42	796	30231	0	0	796	0	1591
43	1591	15115	0	0	0	0	0
44	2387	11933	0	0	0	0	796

Table 22. Percentage of sherds representing each phase in 4-meter sample-circle collections

Sample	Early I	Late I	II	III-A	III-B/IV	V	Unknown
1	9	91	0	0	0	0	0
2	5	62	10	0	0	2	21
3	0	93	7	0	0	0	0
4	4	43	4	0	14	18	18
5	7	88	0	0	0	0	5
6	11	78	4	0	0	2	4
7	11	78	3	0	3	0	5
8	8	65	3	0	0	3	22
9	0	93	0	0	0	5	2
10	3	93	0	0	0	3	0
11	7	79	0	0	0	0	14
12	21	79	0	0	0	0	0
13	4	72	4	0	2	2	15
14	3	79	0	0	0	3	14
15	0	75	8	0	0	0	17
16	4	75	4	0	0	4	13
17	3	77	9	0	0	0	11
18	NUMBER		NOT		UTILIZED		
19	NUMBER		NOT		UTILIZED		
20	NUMBER		NOT		UTILIZED		
21	0	83	17	0	0	0	0
22	0	64	6	0	13	0	17
23	0	0	0	0	0	0	0
24	0	0	0	0	0	0	100
25	0	100	0	0	0	0	0
26	NUMBER		NOT		UTILIZED		
27	14	71	0	0	0	0	14
28	0	72	0	0	0	0	28
29	10	77	0	0	0	0	13
30	4	74	0	0	4	0	19
31	5	68	0	0	2	0	16
32	0	0	0	0	0	0	0
33	19	81	0	0	0	0	0
34	0	100	0	0	0	0	0
35	0	71	2	0	5	0	21
36	0	71	10	0	5	0	15
37	6	81	1	0	1	0	10
38	5	89	2	0	0	0	4
39	0	64	18	0	0	0	18
40	0	79	0	0	0	11	11
41	0	75	0	0	0	25	0
42	2	90	0	0	2	0	5
43	10	90	0	0	0	0	0
44	16	79	0	0	0	0	5

2.   Plowed field with low corn (less than one meter),
      with some other vegetation (such as beans and
      squash) possibly present

3.   Plowed field with high corn (over one meter) with
      other vegetation (such as beans and squash)
      possibly present

4.   Weeds, tomatoes, beans in plowed field

5.   Peanuts and other low ground-cover plants

6.   Old plowed field covered with grass

7.   Unplowed field covered with grass and thornbushes

A density value for each collection was obtained by dividing the
collection area, in meters, into the number of sherds recovered from
that area. Division was carried to four decimal places. The density
values given in table 17 are rim sherds per _hectare_. The main con-
cern in presenting the density values is in providing some quantitative
measure for comparison of units within the survey area.

An initial distinction was made between sample-circle collections
and survey collections, depending on the problems being considered. As
the analysis progressed, it was noted that the sample-circle collections
evidently provided a different class of data than did the survey collections. This can be verified by comparing tables 15 through 17 with
tables 20 and 21. There is a difference of several degrees in magnitude of density values. In terms of both density values and field
techniques, therefore, sample collections and survey collections are
not comparable. This point of noncomparability is further verified
when one compares the percentage of sherds representing each phase
in the two forms of collecting. Some sample collections were obtained from areas that were later survey-collected. The percentages
derived from sample collections of certain units can be compared with
the percentages derived from the same unit when the entire locale was
survey-collected (cf. tables 22 and 18). In summary, there is a
marked difference in both density values and percentages of the ceramics
(sherds) representing each phase when comparing the material collected
from the same collection unit using the four-meter-circle sample-
collection technique and that of complete pickup using the survey-
collection technique.

One final point should be mentioned here in regard to the difference among the terms "collection area," "subarea designation," and
"collection unit" as they appear in table 19 under the column entitled

"Name of Variable." "Collection area" refers to terraces and fields
which were identified on the aerial photographs to orient the crew
in the field. Depending on the size of the terrace or field and the
amount of archaeological material on the surface, it might be divided
into smaller "subareas" for surface collection. The collection-
area and subarea designations were thus assigned in the field in a
general progression of numbers and letters. These designations are
shown in figure 4. "Collection unit" refers to the actual position
that a collection area or subarea occupied within the entire survey,
starting with the first unit collected and numbered sequentially to
the last unit collected (collection unit 121). Thus, any given
collection area might have as many as four (and in one case, five)
distinct collection units (see tables 15-18). A total of 70 collection
areas, comprising 121 collection units, were defined for the Xoxocotlán Piedmont Survey.

11. Analysis of Collection Biases

In any enterprise involving a number of different people, there
will naturally be an equal number of personalities, individual perceptions, and differences in experience. The combination of these widely
varying factors becomes important when analyzing material obtained by
surface collection. The goal of the Xoxocotlán surface collection,
especially the survey portion, was to provide a body of data for
the analysis of patterns of artifact distribution. Unfortunately,
patterning in the distribution of archaeological material does not
necessarily result from the inferred distribution of archaeological
classes. A number of factors, resulting from the manner in which the
material was collected, can combine to produce patterns in collections.
Such counterfeit patterning is the consequence of systematic bias in
the collection.

At this point, it is necessary to make a distinction between two
kinds of bias: "systematic" bias and "random" bias. A consistent
patterning of prejudice in a collection may be termed systematic
bias. This type of prejudice affects archaeological material in such
a way that patterns might be the result either of some real variation
in the archaeological classes or of biased patterning produced by the
collection techniques employed. Unless the degree and direction of
the bias can be assessed, any interpretation of the pattern of distributions must be qualified (Robert Dunnell, personal communication).

Random bias is prejudice that appears in an unsystematic manner
throughout a survey. Because its presence is randomized throughout
a collection, this type of bias does not produce patterns in the archaeological material. Inferences made on the distributions of occupational density, then, can be made with fewer qualifications when the
bias is randomized.

In considering the possible biases in the patterning of the
archaeological materials from the Xoxocotlán piedmont, the discussion
will turn to an examination of potential biases operating in the collection procedures employed. Each potential bias will be defined and
its possible effects on the patterning of material will be outlined.
Next, a number of descriptive and inferential statistics as well as
tables will be provided in order to assess to what degree and in
what direction biases are present in the survey. The importance of
various problems will then be discussed in order to interpret the
validity of the patterns of artifact density found by this project in
the Xoxocotlán region. If the biases are systematic and do, in fact,
have some effect on the problem, correction factors will be suggested.
Such correction factors will aid in the final interpretations so that
the results need not be qualified to the extent necessary without
correction for biases.

PERSONNEL BIASES

A potential bias was manifested in the wide range of field
experience of the crew members. Some crew members had no previous
experience in the region, while others had up to three seasons of
fieldwork in the Valley of Oaxaca. However, in the actual field
operations, it was found that the differences in amount of experience
in the region did not affect the results.

Surface-collection procedures were designed to pick up all rim
sherds that could be perceived in a single thorough examination of the
ground surface. Though experience would have some effect on the visual
acuity and perception of ceramics in the soil, the degree of difference
in the actual pickup capabilities of the crew members would be minimal.

Admitting that, in spite of precautions taken to minimize
the effect of experience, some bias might occur, field techniques
were structured to randomize individual differences. With four people
usually comprising a single collecting crew, no one individual was
responsible for picking up an entire collection unit. Thus, any
single sherd collection would be the result of the activities of from
two to five people. In general, then, the contribution of any single
individual would be equalized by the presence of other collectors.

In a few cases, however, an individual was responsible for an entire
sherd collection. These collection units were noted during the analysis
and interpretation of the results, in order to insure that some sort of

bias was not imposed.

Another potentially serious problem related to individual differences is the placement of personnel within the survey region. Though individual differences can be randomized by using the same crew and equalizing individual differences among the crew, the location of collectors consistently in the same part of the survey region can also produce bias. This was particularly important in the Xoxocotlán Survey, where the irrigation canal was the focal point of hypothesis testing. Consistent collecting by a single individual immediately adjacent to the canal could have systematically biased collections. To avoid such a bias, collection areas were randomly assigned to crew members. This can be seen in figure 4, which shows the random placement of the aerial designations and reflects the random collection of these areas by the project. The absence of any patterning in the order in which various areas were collected insured the reduction, if not the complete elimination, of those biases just noted that could affect the survey results.

The last point to be emphasized in regard to individual differences is that the same people were used for the entire survey. This would seem to assure some consistency in collection procedures and comparability between collections. When more than one crew is involved, multiple correlation of the differences between crew members and possible biases in collecting procedures becomes a much more difficult task.

In summarizing the possible personnel biases operating in the Xoxocotlán Project, a pragmatic view has been suggested--that individual differences are indeed a fact of life. In looking at the composition of the crew, the fact that the same individuals collected the entire survey region, the randomizing of location of crew members within the survey region and within collection units, and the fact that in only a few cases did a single individual collect an entire unit, one can conclude that personnel biases did not have a systematic, deleterious effect on the material collected in the survey portion of the Xoxocotlán Project.

SIZE OF COLLECTION AREA

The importance of the size of the collection area as a form of bias lies in the differential skewing of artifact densities which it might introduce into a surface collection. It is related, of course, to

individual perceptions of artifact density and the way in which
field procedures are structured. In order to assess whether the size
of a collection unit affected the density of artifacts in that unit,
simple inspection of a number of statistical values can be made. This
can also be quantified by using inferential statistics (Spearman's
Rank-Order Test). The following discussion, then, will try to answer
the questions: (1) did the size of the collection units have some
effect on patterning of artifact density; and (2) were smaller units
collected more intensively than larger units? If the answer to either
of these questions is yes, then some portion of the patterning of the
artifact density can be attributed to this form of systematic bias.

The first step in trying to assess this sort of bias is inspection
of the density values of small areas. In this case, small areas are
arbitrarily defined as those collection units having an area less than
2,000 square meters. Twenty-two such units are present in the collection
from the Xoxocotlán piedmont and represent 18% of the total number of
collection units in the survey. In terms of the average size of collection units, approximately 4,000 square meters, these small units
fall considerably below the average in terms of their area. If the
size of collection unit does have a direct relationship to the density
of artifacts in a unit, then the rank-ordering of the smallest areas
according to size should also produce a similar ranking of density
values. That is, the smallest unit should have the highest value,
the next smallest the second highest value, and so on.

Table 23 shows the ranking by size of the 22 smallest units, from
the smallest to the largest, the number of sherds for each, and the
total density of sherds for each. The table indicates that there is
no simple direct correlation between the size of a collection unit and
its artifact density; that is, there is no monotonic decrease in sherd
density accompanying the increase in size of unit.

Table 24 represents the rank-ordering of the highest density values
in the collection. In this case, high density is defined as a density
value over 100. Although the mean for the entire collection was 115, this
mean was affected by a few very high total density values. There were
39 units with densities 100 or larger, roughly 30% of the total number
of collection units (121).

Table 24 also shows the collection-unit number, size of collection
unit, total number of sherds, and type of vegetation on the surface of
the collection unit. Inspection of Table 24 indicates a number of
patterns. First, there is no direct rank-order correlation between

Table 23. Rank-ordering by size of collection units below 2000 m²

Collection Unit	Area in m2	Total Number of sherds	Total Density
9	975	21	215
27	1150	39	339
103	1175	4	34
101	1200	7	58
82	1275	60	471
83	1300	15	115
57	1325	0	0
25	1350	33	244
91	1525	32	210
14	1550	2	13
21	1550	21	135
31	1600	3	19
48	1625	13	80
49	1625	0	0
13	1650	10	61
90	1650	63	382
68	1775	140	739
94	1775	95	535
17	1825	18	99
33	1835	29	159
38	1900	6	32
97	1975	39	197

size of collection unit and total sherd density. Out of 39 cases, 13 (33%) are small units and 26 (66%) are larger units. The distribution of small units relative to density is somewhat random, in that they do not form a continuous distribution among the highest values, but appear throughout the column. Small units are somewhat overrepresented relative to the entire collection, in that 13 of 22 or 59% of the small areas are represented in the highest values, as opposed to 26 of 99 or 26% of the larger-sized areas. Similarly, 50% of the 10 densest values come from small areas, even though these areas constitute only 18% of the population. Visual inspection of the distribution of total sherd density indicates that small collection units are somewhat overrepresented, then, in the highest density values. While the bias is not systematic and constant in terms of distribution, there does appear to be some general relationship.

The next step is to assess whether this relationship is statistically significant. Using the Spearman Rank-Order Test, a correlation coefficient of -.1550 is obtained for the relationship between size of unit and total sherd density for the entire collection. This coefficient is significant at the .090 level. This indicates that there does appear to be a general negative relationship between the size of a unit and total density such that the smaller the unit, the greater the likelihood that it would have a higher-than-average density. However, the statistical significance of the relationship is not high enough, i. e., above the .05 level, to justify concluding that a systematic bias exists.

Table 24. Rank-ordering by density of total density values greater than 100

Collection Unit	Area in m2	Total Number of Sherds	Total Density	Vegetation Type
70	3150	343	1089	3
68	1775	140	789	3
94	1775	95	535	2
82	1275	60	471	4
72	2800	117	418	3
95	2725	108	396	2
90	1650	63	382	3
62	4775	178	373	3
74	4975	180	362	2
27	1150	39	339	2
10	2500	75	294	1
84	5950	172	289	3
76	5900	163	276	6
25	1350	33	244	2
92	2925	65	222	3
18	975	21	215	1
75	5050	108	214	3
91	1525	32	210	3
36	2925	60	205	1
2	6325	128	202	3
66	4000	81	202	2
16	3475	69	199	1
97	1975	39	197	3
23	4525	87	192	4
79	4100	78	190	2
71	3450	64	186	3
51	4725	87	184	6
37	3025	55	182	1
9	4100	74	177	1
15	2900	49	169	1
33	1825	29	159	3
69	2050	31	151	3
20	2250	31	138	3
98	4425	61	138	3
21	1550	21	135	3
83	1300	15	115	3
81	2225	24	108	3
42	4600	48	104	2
73	3400	34	100	3

Another possible bias is the size of the collection unit and its effect on number of sherds collected. In order to assess this bias, the total number of sherds per collection must be related to the size of the unit. These two variables show a rank-order correlation of .1921, which is significant at the .035 level. The interesting thing about this relationship is that it is in the opposite direction of the relationship between size and density. That is, the relationship is a positive one, indicating that an increase in the size of a collection unit is accompanied by a concomitant increase in the number of sherds found in the unit. This is what one would intuitively expect to be the case if bias was not operating in favor of collections in small areas. Furthermore, the relationship is above the arbitrary level of significance (.05), indicating that the relationship between size of unit and number of sherds is statistically significant. There appears to be a general case made, then, that size of collection unit does not appear to seriously bias the survey results.

Perhaps the best possible explanation of the apparent statistical relationship between size of unit and density can be derived from a close examination of the way field boundaries were set up. Initial reconnaissance of a potential survey area was made to note general density and locate possible topographic features that would be useful in subdividing a large area. Those areas that showed few artifacts during the preliminary reconnaissance were not subdivided. Table 25 shows these large areas. In this case, a large area is one with an area greater than 7,000 square meters, well over the average size for the collection as a whole. These large collection units are rank-ordered from largest to smallest. Additional information is present in the form of the collection-unit number, total number of sherds, total sherd density, and type of vegetation. The table shows that, in general, the sherd densities for these 14 largest collection units are rather low. Only four units, or 29%, have densities greater than 50, as compared to a collectionwide average of 57%. Furthermore, 73% of the smallest units have values greater than 50. Similarly, only four collection units in table 3 have more than 50 sherds. It should be noted, however, that only four of the smallest units (table 23) also have more than 50 sherds. These differences in sherd values and density values can be accounted for in part by the vegetation types of the units, which will be discussed later. Most of the large collection units have grass or thorns on their surface, as opposed to a predominance of maize or no vegetation at all on the smaller units. Table 25 does show that there

Table 25. Rank-ordering by size of collection units with areas greater than 7000 m²

Collection Unit	Area in m²	Total Number of Sherds	Total Density	Vegetation Type
105	16825	0	0	5
55	11025	9	8	6
12	11000	19	19	7
54	10750	6	6	6
63	10075	95	94	6
1	9950	82	79	1
45	9550	18	19	7
59	9125	16	18	4
115	8275	8	10	6
89	8050	17	21	2
60	7975	58	73	4
64	7525	0	0	6
61	7350	18	24	4
7	7050	59	84	6

is no monotonic pattern in the distribution of the sherds and sherd densities. There is no simple rank-ordered systematic bias apparent in the table. This lack of directly apparent bias seems to reflect the operational decisions that were made in the field. That is, large areas with grass on them, and with few artifacts indicated by preliminary inspection, were not subdivided. Thus there appears to be a correlation between low density and large areas. This relationship need not be viewed necessarily as systematic bias which would seriously affect the entire collection. This point can be substantiated by looking at the relationships between variables in collection units with areas below 7,000 square meters.

Since the apparent correlation between size and density in large areas can be explained by field operations, the next step is to test whether this explanation accounts for the values in the rest of the collection. That is, if one takes the large areas out of consideration, is there still a relationship between size and density? If there is, the relationship cannot be explained by reference to overt field techniques and must be interpreted as systematic bias. Recalculation of the Spearman correlations for the entire collection without taking into account the large areas tends to show that bias is not present in the patterning of the material. Using 107 cases, one obtains a Spearman coefficient between size and total density of -.0400, which is significant at a .683 level. For those units smaller than 7,000 square meters, the Spearman correlation coefficient between total number of sherds and size is .3106, which is significant at the .002 level. These two findings tend to corroborate the statistical results using rank-ordering techniques which were outlined previously. There appears to be no relationship between size and density in the 107 cases smaller than 7,000 square meters, and a strong, significant positive relationship between size of area and total number of sherds.

In summary, it seems justified to argue that a general, though not highly significant, relationship between size of collection unit and total sherd density does not constitute a systematic bias.

TIME CONSIDERATIONS

Patterns in any collection can be a product of changes in interest and perception on the part of crew members during the course of a project. People simply see things in a different way as their experience increases. Motivation is also variable throughout the length of any long-term project. It usually starts out high and declines in some variable manner throughout the time of the project. Interest picks up when unusual artifacts or a large number of artifacts are found. This day-to-day fluctuation of motivation is also reflected on a smaller scale within a single day. People might work slowly in the early morning, pick up speed during the middle portion of the day, and slow down again in the afternoon sun. The last collection of the day might be completed rapidly in order to get home.

These fluctuating levels of interest are represented in figures 19 and 20. Figure 19 shows the possible fluctuation of interest during a single day. Time of day and various landmark activities are compared with level of interest. Figure 20 shows the fluctuation in level of interest throughout the course of a project. Again, various hypothetical landmarks within the collecting procedures are noted (Robert Dunnell, personal communication).

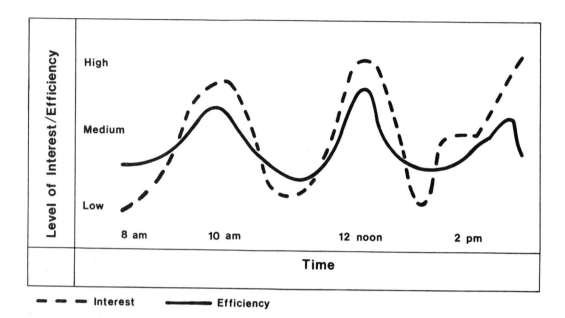

Fig. 19. A schematic model of the fluctuation of interest and collecting efficiency throughout a single day

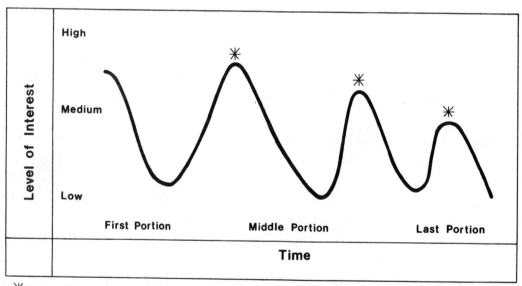

Fig. 20. Schematic model of the fluctuation of level of interest throughout the course of a project

While the fluctuations in interest shown in figures 19 and 20 show
a trend that decreases in efficiency through time, this trend is
counteracted by an increase in experience through time. That is, the
more fieldwork one does in a given area, the more proficient one is at
recognizing artifacts from that region in the local soil types and
various kinds of vegetation. In terms of experience, then, one's
efficiency increases through time. This is shown in figure 21. The
curve shows a gradual increase in efficiency through time. In order
to assess the combination of these effects, one must examine figure 22.
This figure shows the long-term drop in interest level combined with
the increase in experience, which counteracts it to some degree. In
terms of collecting efficiency, then, figure 22 suggests that about
midway through a project the efficiency of most fieldworkers would
begin to drop, as boredom counteracts the increase in experience.
Occasionally the general level of efficiency might rise, when unique
artifacts are found or when large numbers of artifacts are encountered.
This latter problem would have some relationship to sample size and
the number of sherds collected per unit, and will be discussed later.
Once this interest peaks, efficiency once again drops, as fewer sherds
are found in other collection units. The models outlined above, it should
be noted, are just that--models. Actual field experience and information
from other projects indicates these general patterns to be common to
most projects using surface-collection techniques (Robert Dunnell,
personal communication).

The problem to be examined here, then, is whether the chronological
position of a collection unit relative to the length of the project has
any effect on the density. That is, can the patterns of density which
are observed somehow be accounted for in terms of increased experience
or changes in level of interest?

The first scale of analysis is that of variation in collections
made in a single day. Not enough information is presently available to
rigorously assess effects of such variation. All that can be noted is
that no systematic pattern was present in choosing which areas to
collect during any single time of day. Units collected near the end of
the survey were gathered during the optimal time of day for sustaining
interest, usually during the middle of the day. Also, surface collecting
was carried out for only a half day or less at a time. This served to
alleviate boredom, which was increasing near the end of the survey,
after the crew had been working for over a month. There is no reason

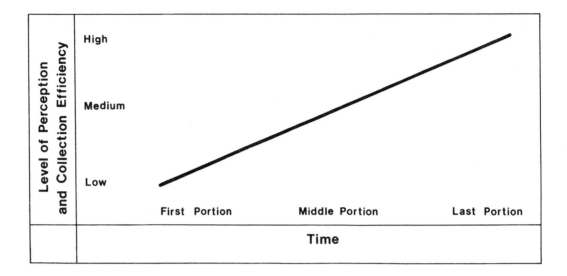

Fig. 21. Schematic model of the change in level of experience and level of perception of artifacts throughout the course of a project

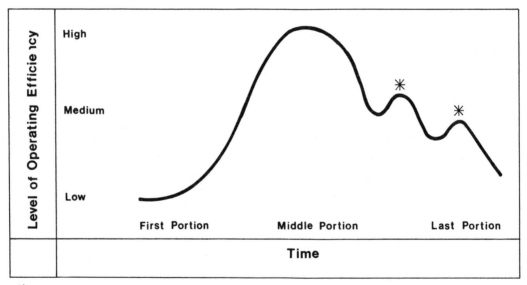

* Unit with many artifacts.

Fig. 22. Overall combination of level of interest and level of experience to form actual in-field operating efficiency during the course of a project

to suspect a systematic bias operating in the collections that could be attributed to the time of day a unit was collected.

In spite of this trend toward randomizing, order of collection units could possibly have introduced bias in another fashion. Collections were made in a general progression along the north side of the canal in an easterly direction, then along the south side of the canal in a westerly direction. This was the case for the last two-thirds of the survey, as can be ascertained by looking at figure 4, which gives the location of collection areas within the survey region. As noted previously, collection areas and subarea designations were then translated into the collection-unit designations, which more directly reflect the relative time a unit was collected within the survey. The time a unit was collected can be ascertained by reference to its collection-unit designation. Figure 4 shows that placement of collection units was generally random within the region. Three major blocks of the region were surveyed during different portions of the survey. Thus, the first block (Areas 1 to 26) was surveyed in the first part of the survey, the second block (Areas 27 to 50) collected in the middle portion, and the last block (Areas 51 to 70) collected in the last few days of the project. The question then arises as to whether the position of a unit within the entire region somehow affects its density. The following discussion will attempt to answer this question.

If there was a direct and obvious relationship between the time when a unit was collected and the density of that unit, a rank-ordering of the results in terms of relative position and of density should show such a pattern. Tables 15 through 17 relate to the first possible bias. Inspection of these tables indicates that there is no structure in the relationship between collection-unit number and density of total sherds. The high densities occur randomly throughout tables 15-17 without forming a unidirectional pattern of distribution. Table 24 rank-orders the high density values according to total sherd density. Again, this table fails to show a direct, patterned relationship between collection unit and total sherd density. Of the 39 highest densities, 23 are in units with numbers 60 and above. That is, 58% of the high density values come from the last half of the collections. Similarly, nine of the 10 densest values come from units above 60. Of interest, however, is that none of the high density values are from the last 23 units collected. If the effect of experience were important and systematic, one would expect these last units to have high values.

Such is not the case. Can boredom have some relationship to the low density values of the last units collected? Since the field operations were structured to alleviate boredom, one can assume that this possibility would be minimized. Thus, the patterning in the results does not appear to be the product of boredom or experience.

The point made above in regard to the absence of systematic bias due to experience and boredom can be supported by other data. Using table 26 one can examine the relationship between low densities and collection-unit number. The purpose of this table is to see if areas with low density are somehow related to time of collection. If so, then special qualification of the density patterning would be necessary when discussing where low densities occurred. In addition to information on collection-unit designation and total sherd density, table 26 notes the total number of sherds and type of vegetation on the collection unit. In this case, low density values are defined as total sherd densities less than 20. These densities are rank-ordered from smallest to largest. Table 26 fails to show a continuous, monotonic increase in collection-unit numbers accompanying the increase in sherd density. Thirteen of these lowest values come from the first half of the survey, and six of the values are from the last 23 collection units. If experience were having a severe influence on density figures, these last units should have much higher values; in fact, they are represented much more abundantly than would be expected in these low density values. Can these differences be attributed to level of interest or boredom? Once again, field techniques were structured so as to hold boredom constant in the last part of the project.

The general pattern of relationships between size and time of collection that is noted by visual inspection can also be analyzed statistically. The Spearman Test shows a relationship between the collection-unit number and total sherd density that has a value of -.1285, which is significant at the .161 level. A rank-order test can be used, since the collection-unit number shows time-ranking according to relative position within the entire collection. In looking at the relationship between collection unit and total density, there does not appear to be a strong statistical correlation. If any relationship is implied, the direction of the relationship is negative. This would indicate that more higher density values were present near the middle or end of the project than would be expected. Also of interest in this regard is the relationship between collection-unit number and total number of sherds for the entire collection. The Spearman Test

shows a relationship between collection unit and total number of sherds
as having a value of -.1800, significant at a .049 level. This indicates a generally strong negative statistical relationship between the
time a unit was collected and the number of sherds found in the unit.
The negative direction indicates that more sherds were found in the
middle or end of the project than would be expected if the number of
sherds were randomly distributed throughout the collection, and this
tends to corroborate the impressions given by the tables in this section.
The important question, then, is: does the relationship between time
and number of sherds constitute a systematic bias in the collection?

Analysis of all evidence presented to this point would tend to
indicate that such a relationship is not in fact a systematic bias.
Further information can be used to bolster this point. Table 27
presents levels of significance and correlation coefficients for
collection-unit number and sherd density, and collection-unit number
and total number of sherds. Coefficients are given for each portion of
the survey. That is, the first block of collection units (1 to 45)
represents the first part of the collection, the second block (46 to 89)
represents the middle portion, and the last block (89 to 121) represents
the last period of time. The table compares the relationships of the
variables to determine whether or not there are differences that might
be attributed to bias in terms of time of collection. All relationships
are compared with the total collection in order to note possible
relationships. In analyzing the relationship between collection number
and total density, one can note a marked degree of variance in strength
and direction of the relationships. A similar difference in patterning
is apparent in relationships between collection-unit number and total
number of sherds. The problem arises of how to account for these differences. Since survey techniques were altered to control bias in the last
portion of the survey, these values can be assumed to reflect actual patterning of the archaeological classes. Since there is no definite patterning for the first part of the survey that is statistically significant, the direction of the relationship is negative. This means that
more sherds were found in the units collected near the end of the first
part of the survey and this is reflected in the density values as well.

Patterns for the second portion of the survey, though significant
in the opposite direction, tend to confirm patterning of the first part
of the survey. There appears to be a positive relationship between
number of sherds and value of the collection-unit number of the second
portion of the survey, though the strength of the relationship does not

132

Table 26. Rank-ordering by density of total density values below 20

Collection Unit	Total No. of Sherds	Total Density	Vegetation Type
8	0	0	6
11	0	0	6
40	0	0	7
49	0	0	5
57	0	0	6
64	0	0	6
105	0	0	5
110	0	0	6
54	6	6	6
58	4	6	6
55	9	8	6
108	2	8	6
115	8	10	5
114	6	15	5
39	4	17	7
59	16	18	4
12	19	19	7
31	3	19	7
45	18	19	7
113	7	20	5

Table 27. Comparison of significance levels and direction of Spearman Correlation Coefficients for the relationship between collection-unit number and total density/total number of sherds

	Total Density with Collection Unit		Total Number of Sherds with Collection Unit	
First Portion of Survey	N=44 .487	–	N=44 .255	–
Middle Portion of Survey	N=43 .012	+	N=43 .161	+
Last Portion of Survey	N=32 .001	–	N=32 .001	–
Values for Entire Collection	N=121 .161	–	N=121 .049	–

The – and + indicate direction of relationship.

appear to be statistically significant. The relationship between collection number and total sherd density, however, does appear to be highly significant statistically. The direction, too, is a positive one, indicating that more material was found in the first portion of the second phase of the collection. This is a continuation of the high values noted in the last part of the first phase of the survey project. Thus, the statistical relationships between time of collection and amount of material found in the first two phases of the project support one another. A large quantity of material was found in the last part of the first phase of collection, and this quantity continued to be found in the first part of the second phase of the project. Since this would be the optimum time for finding material, in terms of the relationship between interest and experience, it is interesting to note that the highest values peak approximately in the middle-to-latter portions of this period in the project. The statistical relationships serve to indicate that the patterning in the archaeological material is not due to bias introduced in terms of interest or experience. There seems to be more material found because there was simply more material in that portion of the region surveyed at the time in question.

A further bit of evidence tends to confirm this observation. The first collection units of the last phase of collection are adjacent to units collected during the second portion of the project. Most units in the general area have high densities. The point to be made is that the units collected in the last phase of the project were collected long after those in the second part of the project, and all have similar density values, regardless of the time they were collected. Thus, the high values obtained by the last collections tend to corroborate the trend established by the collections obtained during the middle of the project. This would seem to indicate the lack of a biased patterning due to interest or experience.

In summary, then, visual and statistical analysis of collection densities and number of sherds collected has failed to suggest a serious, regular, systematic bias that could be attributed to the time within the project a unit was collected. Other problems related to the variation in motivation within a single day have not been examined due to lack of data.

VEGETATION TYPES

As noted previously, modern use of the Xoxocotlán piedmont is relat-

ed primarily to farming. Numerous small plots are worked at different
times of the year depending on such factors as owner, fallow cycle,
type of crop, and season. The importance of this widely ranging
variable can be appreciated when trying to compare collection units in
terms of the condition of the surface at time of collection. The
Xoxocotlán Survey encountered vegetation ranging from grass that covered
100% of the surface area of a collection unit to 20-centimeters-high corn
covering less than 10% of the surface. Other important factors that
affected the number of sherds that could be perceived include soil type,
amount of rain since plowing, and other factors dealing with erosion.
In an ideal situation, one would want to assure a comparable collecting
surface for all units, much in the same manner that field techniques
can be structured to assure that comparable amounts of collection units
are surveyed. Unfortunately, enough information was not available to
assure comparability of the surfaces of all collection units. Since
the parcels of land were being managed by a number of interrelated
factors, there was no means of structuring the survey to assure comparability of collection units in terms of type of surface. Because
of the ejido system and different fallow cycles, for example, it was
not possible to predict when, if ever, a given parcel of grassland would
be plowed. The survey thus proceeded in its collection activities without regard for type of vegetation covering the collection units. During the course of the project, in-field analysis of the sherd density
of several types of vegetation were noticed to vary. This difference in
density values was used pragmatically to set up collection-unit boundaries, as noted above. There does appear to be some justification in
suggesting that many of the low density values in certain vegetation
types are not simply an artifact of collecting procedures. The only
means of adequately testing this statement, of course, is to recollect
areas after they have been plowed. In lieu of such tests, the data
can be made more useful by applying correction factors that attempt to
compensate for systematic biases. Keeping in mind this introduction
to the kinds of problems involved, the discussion will turn to an
assessment of the degree and direction of possible biases introduced
by differences in type of vegetation.

Examination of tables 24 and 26 shows the following pattern: most
of the highest ceramic densities come from vegetation types 1-3, while
most of the lowest densities come from collections with vegetation
types 5-7 (see chapter X for vegetation types). That is, 82% of the
highest densities come from units that had just been plowed and that

contained low or high corn. On the other hand, 90% of the lowest
density values come from units with grass or thorns covering their
surfaces. This general relationship can be examined for the entire
collection to see if the type of vegetation biases the density throughout the entire collection, as well as in the low- and high-density
areas; this is shown in table 28. This table compares the mean density
values for each phase or for the total density relative to the vegetation typology outlined in chapter X. Several patterns are apparent.
First, vegetation types 1-4 have a higher average than the density
values of vegetation types 5-7. The direction of this marked difference
between vegetation types is constant throughout all phases. The
difference between phases varies, however. This can be attributed,
perhaps, to the size of the sample. Taken in conjunction with all the
other tables presented so far, table 28 provides conclusive evidence
that there is a clear systematic bias in the Xoxocotlán Survey collections due to vegetation type.

Table 28. Comparison of density values by type of vegetation

TYPE OF VEGETATION

		1	2	3	4	5	6	7
		N=11	N=34	N=23	N=9	N=8	N=26	N=10
	Total Density	154.455	112.500	233.043	113.111	20.500	58.385	33.600
A V E R A G E D E N S I T Y V A L U E S	Early I	17.909	5.235	16.261	6.556	0.000	3.269	1.900
	Late I	83.182	54.147	125.348	64.000	6.125	26.962	12.600
	II	5.091	7.735	11.174	9.667	0.750	5.077	1.800
	III-A	7.000	0.824	1.826	2.222	0.000	0.423	0.600
	III-B/IV	1.909	3.000	4.826	1.889	0.000	1.192	3.700
	V	4.545	10.824	6.043	9.667	4.875	3.231	1.600
	Unknown	34.636	30.853	67.870	19.111	8.625	18.346	13.600

The density values represent the mean density values for each phase or total
density for the collection. The means are calculated for each type of vegetation based on the number of collection units with each vegetation type;
thus the number of cases per type of vegetation is variable, as indicated
directly below each vegetation type.

This pattern is duplicated by statistical analysis. The Spearman
Test shows a rank-order correlation coefficient of -.3970 between type

of vegetation and total sherd density, which is significant at the
.001 level. The same relationship holds between number of sherds and
type of vegetation. There is a correlation coefficient of -.2499
between total number of sherds and type of vegetation, significant at
the .006 level. Both relationships are in a negative direction, which
suggests that more sherds and higher densities are correlated with the
vegetation types with lowest numbers.

It should be noted that the pattern does not hold for all phases
used in the analysis. For example, table 28 shows that phases with
high densities are more severely skewed by the type of vegetation. In
phases with low densities, on the average, the magnitude of difference
between mean density values is much less. This seems to indicate that
those phases appearing infrequently in the total collection were less
likely to be affected by vegetation than those phases with a lot of
material. This observation is borne out by noting the relationship
between type of vegetation and density by phase. The Spearman Test
shows statistically significant correlations between vegetation and
density values of Early Monte Albán I, Late Monte Albán I, and Unknown
phase (unidentifiable) sherds. These three classes of material have
the densest values throughout the collection. Monte Albán II, Monte
Albán III-A, Monte Albán III-B/IV, and Monte Albán V, all occur
infrequently in the total collection. The relationship between their
density values and type of vegetation is not statistically significant.
This same pattern is duplicated when the number of sherds for each of
these later phases is compared with each type of vegetation. There is
no statistically significant relationship between number of sherds and
type of vegetation for these phases. They simply do not appear frequently enough throughout the total population to be related to any of the
factors related to vegetation. Table 28 does show, however, that there
are some marked differences between vegetation types and mean density
values for at least some of these poorly represented phases.

In summary, analysis of the relationship between vegetation and
density of sherds has shown that, in some cases, a systematic bias is
present. This bias affects those archaeological classes which are the
most abundant in the survey population. This is particularly true when
examining Total Sherd Density, Density of Early Monte Albán I, Late
Monte Albán I, and Unidentifiable Sherds. Though a trend is present in
other classes, the bias for less frequently occurring classes is less
severe and not statistically significant.

The seriousness of the problem can be appreciated by noting that it affects the total number of sherds and total sherd density, heretofore used as general indicators of survey efficiency. The problem is also compounded by its effect on the earliest phases in the survey region. These phases were of primary interest in evaluating hypotheses about occupation relative to the canal. Unless biased results are severely qualified, interpretations based on patterning of occupational density in the Xoxocotlán piedmont must be suspect. In order to alleviate this problem, correction factors can be calculated in an attempt to assure comparability of all archaeological classes. The following discussion outlines the steps taken to formulate correction factors.

In any consideration of effect of vegetation on perception of artifacts in the field, two important controlling variables must be considered. The first is type of vegetation. This is important because it is conditioned by a number of underlying factors, such as depth of soil, amount and kind of erosion, type of soil, and other variables which contribute to plant growth. All of these factors are important in determining what is perceived by a fieldworker. The second controlling factor is percent of ground surface visible, usually expressed as percent of ground cover. This can be the same as or different from type of vegetation, depending on planting techniques, type of crop, spacing of rows, and time since plowing. For example, if time since plowing is rather long, a number of different weeds can spring up in spaces between row crops and severely obscure the ground surface. Thus, it is important to note not only type of vegetation but also the percent of ground cover, since both interact to affect the amount of surface that is visible in the field.

As part of the survey forms of the Xoxocotlán Project, type of vegetation was noted for every collection unit. Thus, this type of information has been coded according to the schema outlined in the previous section. Unfortunately, percent of ground cover was not taken into account while in the field. It can only be approximated by reconstructing its general characteristics based on the information on type of vegetation. Thus, areas with grass on the surface are inferred to have at least 90% of the potentially visible ground surface covered by vegetation, while corn less than 20 centimeters high would obscure less than 10%, depending on the growth of weeds.

The type of vegetation and approximate percent of ground cover, then, are used to derive correction factors for the Xoxocotlán surface material. Correction factors are based on the mean density values for each phase and each vegetation type. The calculations were performed in the following

manner. First, there were no factors calculated for vegetation type 1,
plowed fields without vegetation. This situation was taken as the
optimum condition possible, since 100% of the surface would theoretically be visible. In actual practice, however, one should realize that
even within this situation there is much variability. For instance, a
freshly plowed field presents quite a different surface from a plowed
field that has been rained on. For the purposes of this analysis, however, the class of plowed fields will not be subdivided and will be used
as the standard by which all other vegetation types are compared. If
the mean density of a given phase and a given vegetation type exceeded
the mean density value of that phase for vegetation type 1, then no
correction factor was calculated based on density. If the density value
for a given vegetation type exceeded the optimum, then it was felt that
the pattern actually represented the density of that phase. If, on the
other hand, the density value of a given phase with a given vegetation
type fell below the mean value of that phase with vegetation type 1,
then the lowest value was divided into the highest value to obtain the
relationship in terms of a ratio variable. In table 28, for example, the
mean density value for Late Monte Albán I at optimal conditions (vegetation type 1) is 83.182. The mean density value of Late Monte Albán I
with vegetation type 7 is 12.600. In this case, then, the value for
type 7 would be divided into type 1. The result is a value of 6.601.
This value takes into account the discrepancy between density values,
assuming that the values are randomly distributed throughout the vegetation types. This is not the case, as the different number of units per
vegetation type shows. Nevertheless, the figure does give some sort of
objective measure of the differences between optimal conditions and
different types of vegetation.

     After having calculated the differences between means below the
optimal level on the basis of type of vegetation, one must still take
into account the potential effects of ground cover. That is, even if
a given vegetation type has a mean density average higher than the
optimal conditions, as is the case for most values in vegetation type 3,
there still was some bias operating in the field in the form of percent
of ground cover. That is, even though mean values of vegetation type 3
are higher on the average than values in fields without vegetation, there
is a good possibility that the values would have been even higher if
ground cover had not somehow obscured part of the surface. This seems
entirely likely in the example cited above, since one is talking about
effects of corn almost two meters high. There clearly are different

perceptual factors operating when one flails through head-high corn than when one is walking across a plowed field without any vegetation. In order to take into account these perceptual differences, the percent-of-ground-cover value can be used to further refine the correction factor. This was done in the following way. For those vegetation types having means above the optimal level, the density value was multiplied by a standardized, estimated percent-of-ground-cover value. This figure was then added to the density value to make it higher. Thus, a value of 100 for vegetation type 3 would be multiplied by the standardized estimate of percent of ground cover for that vegetation type (.35), and this value would in turn be added to the original value. Thus, 100 x .35 = 35, 100 + 35 = 135. The final corrected density value for the unit would be 135. Where density values of a given vegetation type were lower than the optimal density, the percent of ground cover was applied to the value obtained by dividing the low mean into the optimal mean. For example, if vegetation type 7 divided into vegetation type 1 produced a value of 1.000, this latter number would be multiplied by percent of ground cover in order to obtain an overall correction factor. Thus, 1.000 x .90 = .9, 1.000 + .9 = 1.900. In this case, then, a correction factor of 1.9 would be multiplied against all values of the phase which occurred in type 7 vegetation. The correction factor is different for each phase and each type of vegetation.

As noted above, survey notes do not include the actual percentage of ground cover for each collection unit, but only the type of vegetation. In order to alleviate this problem, a standardized subjective estimate was made for each major class of vegetation. The estimates that were made represent a median-value estimate of the percent of ground cover. Thus, for any given collection unit the estimated ground-cover value might not be exactly correct. By using the estimated median value, however, the range of variability will be minimized. The following estimated ground-cover values were used for each vegetation type:

Vegetation Type	Ground-Cover Value
1	.00
2	.20
3	.35
4	.25
5	.30
6	.90
7	.90

Certain stipulations about the corrected density values should be

140

made. The values represent the maximum values possible assuming the
maximum possible bias of any vegetation type and percent of ground
cover. The values for any given unit should be interpreted with this
in mind. Obviously, in some cases, the biases were not operating, and
the values reflect unbiased patterning. However, the statistical
relationship throughout the collection shows a general trend in the
pattern of biases. The correction factors are based on the trend for
the collection as a whole. The important point to consider here,
then, is the standardization process. The technique of calculating
correction factors was constant for every case. The justification for
corrections rests in the statistical base, as evidenced in table 28 and
in the statistical relationships derived from the Spearman Rank-Order
Test. The process is thus an objective, statistically justified method,
with all cases treated in the same objective fashion.

The overall purpose of the correction factors is to indicate areas
in the collection region where possible vegetation biases could have
been operating. The corrected density-value distributions can be compared with uncorrected distributions to see if the same patterning of
archaeological classes holds. If there are, in fact, differences between
the two sets of distributions, then vegetation differences must be taken
into account, along with other factors, in order to assess the validity
of patterning. The point is, as long as one has corrected patterns
determined in an objective manner, then such comparisons can be made.

SAMPLE SIZE

As is evident by inspecting tables 15 and 16, there is a marked
contrast in the number of sherds collected for different phases. The
most underrepresented phases appear to be Monte Albán II, Monte Albán
III-A, Monte Albán III-B/IV, and Monte Albán V. Since the number of
sherds for each of these phases is small compared to Late Monte Albán
I, there is a good chance that some sort of bias affects these phases
more intensely than the better-represented phases. In order to discuss
patterning of such low numbers of sherds, severe qualifications of the
interpretations must be made. Nevertheless, the fact that certain phases
do have differing patterns of distribution could potentially be important. This section will attempt to establish relationships between size
of the small samples and potential biases such that some of the poorly
represented phases may shed light on the overall pattern of distribution
in the Xoxocotlán piedmont.

   Perhaps the most important bias operating on small sample sizes is
the change in interest levels. In the Xoxocotlán region, where some
phases are obviously poorly represented in the sample, this bias might
be present. For example, standard survey techniques previously described
are designed to obtain most of the surface material. If a member of
a population is present in low frequencies relative to the population
at large, chances are good that this component of the population will show
up rather infrequently in sherd counts. However, a change in interest,
as when a certain collection suddenly has a high density of a well-
represented phase, increases the likelihood of the low-density phase
being collected. Thus, it is possible that the low-density phases might
achieve their highest levels as the result of some change in interest.
The patterning that would be apparent, then, would show a phase with
low density having the same nodes of high density values as the best-
represented phase in the collection. The patterning might not be due
to the fact that the high-density node was occupied throughout a number
of different phases, but rather to a change in collecting biases which
skewed the density values of the low samples. In this survey, there are
obviously more Late Monte Albán I phase sherds than all others combined.
Some of the less-well-represented phases appear to have distributions
similar to the material from Late I. The question then arises, how
valid are the patterns for Monte Albán II, III-A, III-B/IV, and V?
Could their highest values be attributed to interest? In order to
examine this problem, the collection units with the highest number of
sherds can be tabulated and the pattern of sherd distribution for
these units outlined.

   Table 29 shows the rank-ordering of collection units on the basis
of total number of sherds. Those units with more than 50 total sherds
are ranked from largest total sherd collection to smallest collection.
Number of sherds for each phase is indicated. This table confirms the
fact that Late Monte Albán I is by far the best-represented phase in
all the survey collections. There is also a large number of sherds
which could not be identified as to phase. Judging from the pattern
in the collections, though, one would estimate that a large proportion
of these unidentified sherds would be identified as Late Monte Albán
I if they had not been eroded, etc. The highest sherd numbers are
distributed throughout the collection, but a larger proportion of the
collection units occur in the middle of the second portion of the survey. This verifies, generally, the pattern of sherd density outlined
previously. Those phases with few sherds appear to have some of the

Table 29. Rank-ordering by total number of sherds of units with more than 50 sherds

Collection Unit	Total Sherds	Early I	Late I	II	III-A	III-B/IV	V	Unknown
70	343	32	212	6	0	11	0	82
74	180	7	96	19	0	1	0	57
62	178	9	105	11	0	0	1	52
84	172	7	98	15	0	0	10	42
76	163	9	102	11	2	0	7	32
68	140	7	64	5	6	8	0	50
2	128	9	72	8	3	0	2	34
72	117	12	74	2	0	0	1	28
75	108	6	65	1	1	5	0	30
95	108	7	64	14	0	0	2	21
63	95	8	51	9	0	0	6	21
94	95	3	45	7	0	0	15	25
23	87	10	35	5	3	4	8	22
51	87	8	43	6	0	0	4	26
1	82	12	43	6	0	0	2	19
66	81	6	41	6	0	3	0	25
79	78	2	35	6	1	0	13	21
10	75	4	46	4	2	0	0	19
9	74	5	66	3	0	0	0	0
16	69	8	32	1	4	0	0	24
92	85	4	32	6	0	1	5	18
71	64	3	35	0	0	0	0	26
90	63	4	32	3	0	0	2	22
98	61	5	38	0	0	0	3	15
36	60	6	37	4	0	0	0	13
82	60	2	44	7	0	0	7	0
7	59	1	28	7	0	0	0	23
60	58	5	31	2	0	0	2	18
37	55	5	30	3	0	1	0	16

highest sherd numbers randomly scattered throughout the table, except for Monte Albán II, which has virtually all of its highest sherd counts appearing in the collection units with the highest total sherd counts. The number of unidentified sherds remains rather constant throughout all collection units. The fact that there are so many units without these poorly represented phases, even though the table shows the highest sherd counts, would tend to indicate that there is no obvious relationship between the high total sherd counts and the presence of underrepresented phases. However, in order to more fully comprehend the relationship, some comparison must be made of the location of the highest values of Monte Albán II, III-A, III-B/IV, and V. If the highest sherd counts of these phases correlate with high total sherd counts, then one could suggest that interest levels played an important role in the distribution of these phases.

Table 30 presents the rank-ordering of a number of sherds from
Monte Albán II, Monte Albán III-A, Monte Albán III-B/IV, and Monte
Albán V. Collection units with the highest 10 or so sherd counts for
each of these phases have been rank-ordered from highest values to
lowest. Values in this table can be compared with those in table 29
in order to ascertain the relationship between high total sherd counts
and high values for each of these phases. One can thus note the number of shared collection units in table 29 and table 30 in order to
determine whether total number of sherds can account for the high sherd
values of these phases. If there is a relationship, then the high
values can probably be accounted for by changes in interest. That is,
if crew members find a lot of material in a collection unit, they will

Table 30. Rank-ordering of units by highest number of sherds for Monte Albán II,
Monte Albán III-A, Monte Albán III-B/IV, and Monte Albán V

Coll Unit	MA II	Coll Unit	MA III-A	Coll Unit	MA III-B/IV	Coll Unit	MA V
74	19	17	6	70	11	94	15
84	15	68	6	99	9	79	13
95	14	15	4	68	8	84	10
62	11	16	4	100	7	89	11
76	11	2	3	102	7	106	9
63	9	22	3	41	6	23	8
2	8	23	3	75	5	78	8
7	7	47	3	23	4	76	7
82	7	10	2	27	4	82	7
94	7	27	2	45	4	63	6
1	6	46	2			15	5
51	6	76	2			25	5
66	6					56	5
70	6					77	5
79	6					91	5
92	6					92	5
93	6					104	5
						51	4
						80	4
						86	4
						107	4
						120	4

unconsciously collect more material than they might have otherwise. This extra material could quite possibly be sherds from poorly represented phases that would normally not appear in the collection.

In looking at the first columns in table 30, one can compare the number of collection units with high Monte Albán II sherd counts with the 29 units of high total sherd counts in table 29. Of the 17 units with relatively high Monte Albán II sherd counts (in this case, high sherd counts for each phase are determined relative to each phase; thus more than five sherds is a relatively high value for Monte Albán II), 15 of these units or 88% also occur in the 29 units with the highest total sherd counts. For Monte Albán III-A, six of 12 (50%) of the highest sherd counts appear in the same units as in table 29. The highest sherd count for this phase is only six. Monte Albán III-B/IV has a maximum count of 11 sherds per unit. Four of the 10 highest (40%) are listed in table 29. Finally, the highest count per unit of Monte Albán V sherds is 15 and eight of the 22 collection units with the highest Monte Albán V counts (36%) appear in table 29.

In terms of the differential effects of sample size and interest, Monte Albán II and Monte Albán III-A appear to be related to high sherd counts for total number of sherds. Monte Albán III-B/IV appears to be somewhat less conditioned, but it has so few sherds in any case that there still is a possibility that high total sherd counts might somehow affect the density values for this phase. The best case for lack of bias appears to be the distribution of Monte Albán V sherds. This phase shares only 36% of the collection units with the highest sherd counts and also has more sherds overall than Monte Albán III-A and III-B/IV.

With this subjective assessment in mind, one can turn to the Spearman correlations for the relationships that were visually outlined in tables 29 and 30. There is a statistically significant relationship between total number of sherds and the number of sherds for all the phases discussed above. All significance levels are better than .05, although the values for III-B/IV and V are not as great as the other phases. The relationship also holds for total density and density of the underrepresented phases. All values between total density and the density of these phases are above the .05 level of significance, although once again the levels of significance are not as great for III-B/IV. These findings indicate a bias operating in the distribution of Monte Albán II, III-A, III-B/IV, and V, or at least the possibility that biases are operating.

In summary, interpretation of the distribution of phases in the
Xoxocotlán piedmont that occur infrequently indicates a possible bias
relative to changes in levels of interest in dense collections. This
bias could possibly affect the density levels of phases with few sherds.
In order to interpret levels of high density of these phases, it is
suggested that each density node be evaluated relative to the total
number of sherds in the collection units with high densities. If
high-density nodes do correspond to collections with high total sherd
counts, then these apparent high-density levels can be interpreted
as artifacts of the collecting procedures. If, on the other hand,
these nodes do not correspond to units which might have been collected
more intensively, then the patterning in the distribution need not be
attributed to this type of bias.

SUMMARY

The preceding section has examined material from survey collections
in the Xoxocotlán piedmont in regard to systematic and random bias. A
number of potential systematic biases were suggested, and each in turn
was defined and tested by means of descriptive and inferential statistics. Visual and statistical analysis indicated that, in some cases,
systematic biases were indeed operating in the surface-collecting
procedures. Patterning in distribution of artifact densities in these
biased cases must be qualified in order to take into account the effect
of bias. The most serious systematic bias throughout the collection is
type of vegetation. A means of attenuating the effects of this bias
was suggested in the form of a correction factor. Other possible
systematic biases affecting certain density values include size of a
collection unit and differential effects of interest. This latter
bias can be introduced in a number of ways, but the most serious problem
appears to be related to skewing in very dense total collections.

The point of the section is to suggest a means of interpreting
the distribution of artifact density. With the possible effects of
biases in mind, one can interpret patterning in the distribution of
archaeological material with less qualification.

12. Analysis of Survey Results

The results of the survey are presented on computer density maps
using the SYMAP program. The density of rim sherds per hectare for
each collection area or subarea (collection unit), hereafter referred
to as Areas, is calculated by the computer (see table 15) and these
values are used to make contour maps of sherd density for each phase.
When adjacent Areas differ in value (sherd density), the computer
interpolates and prints intermediate values between them.

Densities are presented on the maps in terms of arbitrarily set
density levels, each of which is represented by a particular map symbol
(fig. 23). Originally, four maps were made for each phase: a Primary
Map, showing rim-sherd density per hectare; a Corrected for Vegetation
Map, with values increased by a correction factor to allow for vegetational differences (see chapter XI); a Levels Reset Map, with the levels adjusted for the particular phase so that the density values are
distributed fairly evenly throughout all levels; and a Corrected for
Vegetation Map with Levels Reset, which has values altered by vegetation and levels reset for the particular phase. To conserve space,
however, some of the maps have been deleted. Those retained are the
Primary Map for each phase (figs. 27, 30, 34, 37, 39, and 41), the
Corrected for Vegetation Map for each phase (figs. 29, 32, 36, 38, 40,
and 42), and the Levels Reset Map for Early and Late Monte Albán I and
Monte Albán II (figs. 28, 31, and 35). In addition, a special map
defining possible residential areas has been included for Late Monte
Albán I (fig. 33). The total density of all sherds collected, the
total density corrected for vegetation, and the density of unidentifiable sherds are shown in figures 24, 25, and 26.

The Primary Maps and the Corrected for Vegetation Maps have the
same density levels as those employed on the map of total sherd density
(fig. 24), so that they are all directly comparable. These levels are

MINIMUM	0.00	10.00	30.00	50.00	80.00	100.00	200.00	300.00	1000.00
MAXIMUM	10.00	30.00	50.00	80.00	100.00	200.00	300.00	1000.00	3500.00
LEVEL	1	2	3	4	5	6	7	8	9

```
================================================================================
........  --------- ========= +++++++++ 000000000 XXXXXXXXX 000000000 ......... .........
........  --------- ========= +++++++++ 000000000 XXXXXXXXX 000000000 ......... .........
SYMBOLS ....1.... ----2---- ====3==== ++++4++++ 000050000 XXXX6XXXX 000070000 ....8.... ....9....
........  --------- ========= +++++++++ 000000000 XXXXXXXXX 000000000 ......... .........
........  --------- ========= +++++++++ 000000000 XXXXXXXXX 000000000 ......... .........
================================================================================
```

Levels and symbols used on Primary and Corrected for Vegetation Map

MINIMUM	0.00	1.00	5.00	10.00	15.00	20.00	25.00
MAXIMUM	1.00	5.00	10.00	15.00	20.00	25.00	105.00
LEVEL	1	2	3	4	5	6	7

```
==================================================================
........  ========= +++++++++ 000000000 XXXXXXXXX ......... .........
........  ========= +++++++++ 000000000 XXXXXXXXX ......... .........
SYMBOLS ....1.... ====2==== ++++3++++ 000040000 XXXX5XXXX ....6.... ....7....
........  ========= +++++++++ 000000000 XXXXXXXXX ......... .........
........  ========= +++++++++ 000000000 XXXXXXXXX ......... .........
==================================================================
```

Levels and symbols used on Early Monte Alban I Levels Reset Map

MINIMUM	0.00	1.00	4.00	10.00	20.00	35.00	75.00	150.00	300.00
MAXIMUM	1.00	4.00	10.00	20.00	35.00	75.00	150.00	300.00	1000.00
LEVEL	1	2	3	4	5	6	7	8	9

```
================================================================================
........  --------- ========= +++++++++ 000000000 XXXXXXXXX 000000000 ......... .........
........  --------- ========= +++++++++ 000000000 XXXXXXXXX 000000000 ......... .........
SYMBOLS ....1.... ----2---- ====3==== ++++4++++ 000050000 XXXX6XXXX 000070000 ....8.... ....9....
........  --------- ========= +++++++++ 000000000 XXXXXXXXX 000000000 ......... .........
........  --------- ========= +++++++++ 000000000 XXXXXXXXX 000000000 ......... .........
================================================================================
```

Levels and symbols used on Late Monte Alban I Levels Reset Map

MINIMUM	0.00	1.00	3.00	9.00	14.00	20.00	30.00
MAXIMUM	1.00	3.00	9.00	14.00	20.00	30.00	75.00
LEVEL	1	2	3	4	5	6	7

```
==================================================================
........  ========= +++++++++ 000000000 XXXXXXXXX ......... .........
........  ========= +++++++++ 000000000 XXXXXXXXX ......... .........
SYMBOLS ....1.... ====2==== ++++3++++ 000040000 XXXX5XXXX ....6.... ....7....
........  ========= +++++++++ 000000000 XXXXXXXXX ......... .........
........  ========= +++++++++ 000000000 XXXXXXXXX ......... .........
==================================================================
```

Levels and symbols used on Monte Alban II Levels Reset Map

MINIMUM	0.00	20.00	40.00	60.00	80.00	100.00	120.00	150.00	200.00
MAXIMUM	20.00	40.00	60.00	80.00	100.00	120.00	150.00	200.00	675.00
LEVEL	1	2	3	4	5	6	7	8	9

```
================================================================================
........  --------- ========= +++++++++ 000000000 XXXXXXXXX ......... ......... .........
........  --------- ========= +++++++++ 000000000 XXXXXXXXX ......... ......... .........
SYMBOLS ....1.... ----2---- ====3==== ++++4++++ 000050000 XXXX6XXXX ....7.... ....8.... ....9....
........  --------- ========= +++++++++ 000000000 XXXXXXXXX ......... ......... .........
........  --------- ========= +++++++++ 000000000 XXXXXXXXX ......... ......... .........
================================================================================
```

Residential Areas defined as having values over 100.(Late Monte Alban I)

Fig. 23. Levels and symbols used on computer density maps

Fig. 24. Total sherd density, all phases, Primary Map

Fig. 25. Total sherd density, all phases, Corrected for Vegetation Map

151

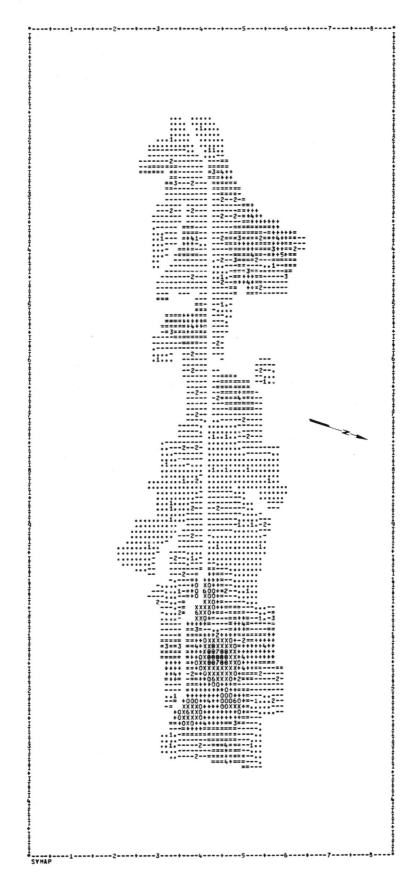

Fig. 26. Density of unidentifiable sherds

given in table 31. The Levels Reset Maps have the levels changed to
fit the particular phase, so that the highest density value, even if
very low, is at least Level 7. This allows use of levels which encompass a smaller range of values, so that finer distinctions can be made
in the distribution of density values, especially in phases with relatively low densities. As discussed in chapter XI, the increased values
used to make the Corrected for Vegetation Maps represent the maximum
amount of error which could have been caused by vegetation. The
actual density value is probably somewhere between the value shown on
the Primary Map and that shown on the Corrected for Vegetation Map.
The discussion of the density maps by phase which follows will con-
centrate on the Primary Maps, but mention will be made of the Levels
Reset Maps and the Corrected for Vegetation Maps where significant
differences occur.

The number of Areas having density values falling within each
level for each map is given in tables 32 and 33, and the percent of
surface area occupied by various levels for each phase is given in
table 34. Table 35 shows the number of Areas having one or more rim
sherds by phase and number of rim sherds collected per phase.

The computer density maps should be compared with the base map of
the area surveyed (fig. 4). All numbered Areas on the base map are
represented by density symbols on the computer maps. Blank areas on
the computer density maps were not collected, and represent steep
eroded surfaces which are sides of arroyos and barrancas, as can be
seen by looking at the drainage pattern on the base map. The blank
line down the center of the computer maps is the route of the ancient
canal, which is traceable on the surface as a shallow depression and
two parallel lines of small stones, probably the result of canal
cleanings. The canal follows the top of the ridge, with land sloping
away from it on both sides, until it reaches Area 37a, where it angles
off to the south and runs below the crest of the ridge until it empties
into an arroyo. Below (east of) Area 37a, the ridge crest runs through
Areas 38a, 39a, 40a, 41a, 42a, 43a, 44a, and 45a. The dam, which is
the origin of the canal, and Monte Albán are off the map to the west
of the map area.

Inspection of the computer density maps will show that sometimes
the number of the level is printed instead of the symbol for the level.
These numbers represent the actual level of the density value of the
Area, whereas the surrounding symbols may be interpolations. The
location of each number is the location, as given to the computer, of

Table 31. Levels used on Primary and Corrected
for Vegetation maps

Level	Density Value
1	0 - 10
2	10 - 30
3	30 - 50
4	50 - 80
5	80 - 100
6	100 - 200
7	200 - 300
8	300 - 900
9	900+

Table 32. Number of collection units per standard density level

Level	1	2	3	4	5	6	7	8	9
Total	12	21	17	21	11	18	11	9	1
Total, Corrected for Vegetation	8	4	11	16	8	25	16	29	4
Early MA I	86	30	4	0	0	1	0	0	0
Early MA I, Corrected for Veg.	58	31	11	13	2	6	0	0	0
Late MA I	32	32	17	11	6	16	4	3	0
Late MA I, Corrected for Veg.	22	22	12	16	7	25	6	11	0
MA II	92	25	2	2	0	0	0	0	0
MA II, Corrected for Vegetation	80	29	9	3	0	0	0	0	0
MA III-A	115	4	2	0	0	0	0	0	0
MA III-A, Corrected for Veg.	100	8	8	0	0	5	0	0	0
MA III-B/IV	111	7	3	0	0	0	0	0	0
MA III-B/IV, Corrected for Veg.	105	12	2	2	0	0	0	0	0
MA V	94	21	4	1	1	0	0	0	0
MA V, Corrected for Vegetation	83	27	8	2	0	1	0	0	0
Unknown	36	46	13	18	1	5	2	0	0

Table 33. Number of collection units per reset density level

Level	1	2	3	4	5	6	7	8	9
Early MA I	54	12	20	11	11	7	6	0	0
Late MA I	15	5	12	21	15	23	17	10	3
MA II	48	7	35	11	8	8	4	0	0

Table 34. Percent by phase of total surface area surveyed (51.26 ha) with densities of Level 2 and above, Level 3 and above, and Level 6 and above

Phase	Level 2+	Level 3+	Level 6+
Early MA I	23.6%	2.3%	0.6%
Late MA I	70.3%	42.7%	12.7%
MA II	18.6%	2.1%	0.0%
MA III-A	2.4%	0.7%	0.0%
MA III-B/IV	5.3%	1.2%	0.0%
MA V	15.6%	2.1%	0.0%

Table 35. Number of areas having one or more rim sherds by phase and number of rim sherds collected per phase

Phase	Number of Areas Having Rim Sherds	Number of Sherds Collected
Early MA I	67	314
Late MA I	106	2365
MA II	73	269
MA III-A	23	51
MA III-B/IV	32	102
MA V	61	226
Total		3327

a specific collection Area. Thus, each Area is represented by a point on the computer maps with specific grid coordinates. In the following discussion of the computer density maps, locations of Areas referred to by number, such as Area 40a, can be located on the computer maps by using the coordinate numbering system found in the computer map borders and table 36, which gives the grid coordinates for each Area shown on the base map.

For descriptive purposes, the survey area will be divided into three sectors. The Upper Sector includes Areas 1 through 27 excepting Area 3. The Middle Sector includes Areas 28 through 36 and 59 through 69 plus Area 3. The Lower Sector comprises Areas 37 through 58 plus Area 70. In terms of computer map coordinates, all Areas between vertical coordinates zero and 58 are in the Upper Sector, the zone between vertical coordinates 58 and 97 is the Middle Sector, and Areas between coordinates 97 and 140 are in the Lower Sector.

EARLY MONTE ALBÁN I PHASE

The Primary Map for Early Monte Albán I (fig. 27) shows one node of fairly high density in the Lower Sector plus a scattering of lesser values in the Upper Sector, while the Middle Sector is almost entirely lacking in Early Monte Albán I material. The node in the Lower Sector centers in Area 40a, with a density value of 102, which is Level 6 (density values between 100 and 200). Areas 39a and 41a, which are immediately adjacent to Area 40a, are Level 3 (30 to 50), with density values of 39 and 43, respectively. This node in Areas 39a, 40a, and 41a is surrounded by a large zone of Areas with Level 2 densities (10 to 30). These are Areas 37a, 38b, 38c, 42a, 42b, 43a, 46a, 47, 51a, 51b, 51c, 52, 53, 54a, and 55b.

The total area of Level 6 density in the Lower Sector is 3,150 square meters. The area of Level 3 is 4,575 square meters, and the area of Level 2 is 55,000 square meters (5.5 hectares). The total area of the Lower Sector is 161,950 square meters (16.2 hectares). Level 6 occupies 1.9% of this area, Level 3 occupies 3.5%, and Level 2 occupies 29.4%. Thus, 38.2% of the Lower Sector is Level 2 or greater.

The Middle Sector has very little Early Monte Albán I material. There are two Areas which are Level 2: Areas 29b and 3a. The total surface area of these two Areas is 7,900 square meters. The total area of the Middle Sector is 191,425 square meters (19.14 hectares). Thus, only 4% of the area of the Middle Sector was Level 2, which was

Table 36. Coordinates of collection units for SYMAP figures

COLL AREA	COLL UNIT	VERT COOR	HORI COOR	COLL AREA	COLL UNIT	VERT COOR	HORI COOR	COLL AREA	COLL UNIT	VERT COOR	HORI COOR
1	1	45	38	24A	41	28	38	45B	81	133	46
2	2	54	38	24B	42	28	34	46A	82	107	54
3A	3	74	43	25A	43	33	38	46B	83	107	57
3B	4	74	46	25B	44	33	33	47	84	112	55
3C	5	80	43	26	45	38	31	48A	85	117	54
3D	6	80	46	27A	46	50	45	48B	86	118	58
4A	7	55	33	27B	47	57	44	49A	87	122	53
4B	8	60	30	28A	48	62	54	49B	88	122	57
5A	9	38	39	28B	49	64	55	50	89	130	52
5B	10	38	38	29A	50	67	44	51A	90	106	39
6	11	20	33	29B	51	67	49	51B	91	109	39
7	12	24	33	30	52	71	50	51C	92	112	39
8A	13	46	45	31	53	74	51	52	93	117	38
8B	14	45	45	32A	54	80	51	53	94	124	37
8C	15	44	51	32B	55	82	56	54A	95	117	34
8D	16	46	51	33A	56	90	55	54B	96	121	34
8E	17	42	53	33B	57	94	54	55A	97	112	35
9	18	40	57	34	58	90	52	55B	98	112	32
10A	19	42	45	35	59	90	49	56	99	106	35
10B	20	42	48	36A	60	87	44	57A	100	102	35
10C	21	41	52	36B	61	94	44	57B	101	104	30
11A	22	38	45	37A	62	102	41	58	102	99	36
11B	23	38	49	37B	63	102	46	59	103	96	34
12	24	38	54	37C	64	102	51	60	104	96	38
13	25	38	58	38A	65	110	44	61	105	94	28
14	26	40	62	38B	66	108	50	62	106	92	38
15	27	41	59	38C	67	111	50	63	107	87	38
16	28	45	60	39A	68	113	44	64A	108	86	33
17	29	43	57	39B	69	114	50	64B	109	87	33
18	30	47	54	40A	70	115	44	64C	110	89	33
19A	31	34	45	40B	71	116	50	65A	111	76	39
19B	32	34	48	41A	72	118	43	65B	112	82	39
19C	33	34	52	41B	73	118	49	66A	113	76	36
20A	34	31	45	42A	74	122	48	66B	114	82	36
20B	35	31	48	42B	75	122	42	67	115	79	32
21A	36	26	43	43A	76	126	41	68A	116	66	39
21B	37	26	45	43B	77	126	48	68B	117	70	39
22A	38	22	43	44A	78	130	40	69A	118	59	38
22B	39	22	42	44B	79	130	46	69B	119	62	38
23	40	18	40	45A	80	132	39	70A	120	128	33
								70B	121	130	33

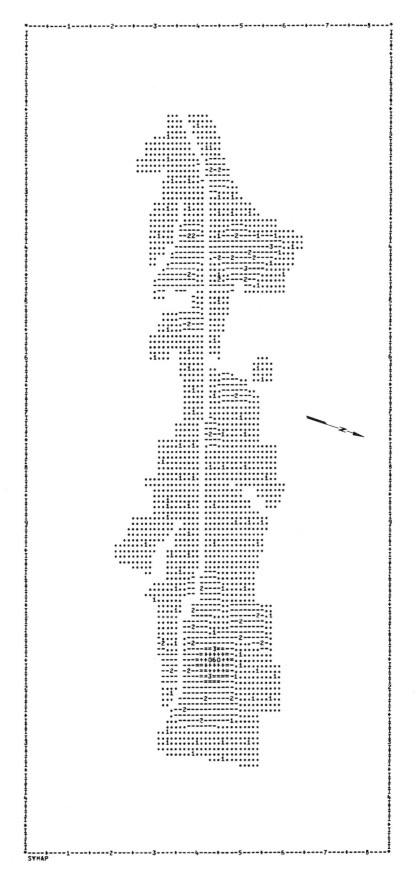

Fig. 27. Early Monte Albán I, Primary Map

the highest density in the Sector.

The Upper Sector has values of Level 3 density in Areas 8c and 9. These are accompanied by Areas 8a, 8d, 8e, 10a, 10b, 10c, and 11b, which are Level 2 and are located between the Level 3 Areas (Areas 8c and 9) and the canal. Areas 21a and 21b are also Level 2, as are Areas 1, 2, and 5b on the south side of the canal. The surface area of Areas with Level 3 densities in the Upper Sector is 3,875 square meters. The surface area with Level 2 densities is 47,050 square meters. The total area of the Upper Sector is 159,240 square meters, so that Level 3 densities occupy 2.4% of the Upper Sector and Level 2 densities occupy 29.6%. Thus, 32% of the Upper Sector is occupied by Level 2 or Level 3 densities.

The total area of all collection units is 512,615 square meters or 51.3 hectares. The total area having densities of Level 2 or greater is 121,550 square meters or 23.6% of the total area. Level 3 or greater densities occupy 11,600 square meters or 2.26% of the area surveyed. Level 6 occupies 3,150 square meters or 0.6%.

The Levels Reset Map for Early Monte Albán I (fig. 28) breaks down the distribution of density values so that approximately equal numbers of cases are assigned to each level. This makes internal variation in the levels used on the Primary Map visible. The Levels Reset Map shows that values between 5 and 25 cluster along the canal, and around the values above 30 which occur in two clusters (Areas 39a, 40a, 41a, and Areas 8c and 9), each about 100 to 200 meters away from the canal. However, it should be noted that whereas all Areas with values over 30 appear as Level 7, four of them have values between 30 and 45, while the fifth, Area 40a, is clearly of a different order of magnitude, with a density value of 102. An exception to the above distribution is the gap in sherd distribution along the canal in Area 36b and Area 62, both of which have density values of zero. These two Areas mark the boundary of what have been termed the Middle and Lower Sectors.

The Corrected for Vegetation Map for Early Monte Albán I (fig. 29) shows that Area 29b may have been as densely occupied as Area 40a. The correction factor for grass increases the density value of Area 29b from 17 to 166, whereas Area 40a with high corn went from 102 to 176. Area 43a, also covered by grass, becomes Level 6; the correction factor increased the density value from 15 to 147. The other Level 6 values occur in Areas 25a and 54a. Area 25a is a thornbush area, where systematic collection was impossible. This condition also applies to

159

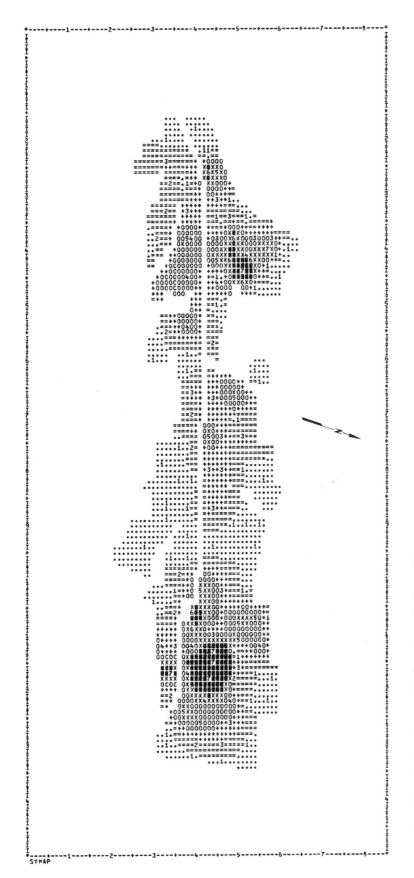

Fig. 28. Early Monte Albán I, Levels Reset Map

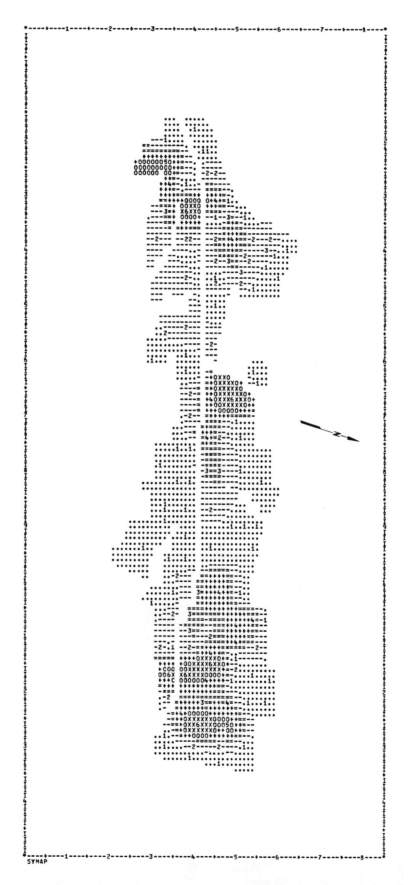

Fig. 29. Early Monte Albán I, Corrected for Vegetation Map

Area 7 and Area 24b, which increased to Levels 5 and 4, respectively.
These bushy Areas were probably overcollected in some more open areas
and not at all in others. The corrected values for bushes (vegetation
type 7) are probably much less reliable than for the other vegetation
types. The other new Level 6 Area is Area 54a, which, somewhat sur-
prisingly, is covered only by low corn. The value for this Area
changed from 26 to 100.

Low corn was also responsible for increasing Areas 3a, 38b, and
38c from Level 2 to Level 4. Areas 11b and 46a were increased to
Level 4 by weeds. Finally, grass in Areas 20a, 29a, and 37b increased
the density values of those Areas to Level 4, and increased the density
value of Area 43b to Level 5.

EARLY MONTE ALBAN I: DISCUSSION

The distribution of Early Monte Albán I sherds shows a definite
orientation with respect to the canal. The distribution of low densities
along the canal (which is readily visible on the Levels Reset Map,
fig. 28) is complemented by nodes of higher density away from the canal
which are surrounded by zones of zero or very low density. This would
indicate a pattern of agricultural use along the canal with residences
located away from the canal, off the best agricultural land (the land
closest to the canal). The major node in Area 40a, in the Lower Sector,
is actually topographically higher than the canal and therefore does
not occupy land that could have been irrigated.

The attribution of agricultural use to land where there is a low
density of sherds over a wide area follows from a practice still in
evidence today: household trash is taken out to the fields for fertilizer. In the past this resulted in the spreading of sherds over the
fields, but now one finds items of plastic (which may be of interest
to future archaeologists). Thus, a low density of sherds along the
canal, with no or very few sherds away from it, may indicate that
land along the canal was used agriculturally, and was receiving water
from the canal and trash from residential areas.

The Primary Map for Early Monte Albán I (fig. 27) shows one node
of much greater magnitude than any of the others. This is Area 40a
in the Lower Sector, which has a value of 102 (Level 6). The next
highest density value is 43 in Area 41a, adjacent to Area 40a. The
highest value away from this node is 31 in Areas 8c and 9 in the Upper
Sector. Thus, if any area is to be considered to have been residential,

it must be Area 40a, which is clearly a nodal Area of much greater
density than anywhere else in the survey area. As noted above, Area
40a is located on land which could not have been irrigated by the canal.

The Corrected for Vegetation Map (fig. 29) shows other possible
nodes of the same magnitude as Area 40a in Areas 25a, 29b, and 43a.
Note that Area 25a is a thornbush Area, which may have skewed results,
while Areas 29b and 43a are grassy Areas. It should be stated again
here that the corrections for vegetation indicate the possible range
of error caused by vegetation. Therefore, the Corrected for Vegetation
values are the maximum densities that may exist under the vegetation.
The actual density is probably somewhere between that given on the
Primary Map (fig. 27) and that given on the Corrected for Vegetation
Map (fig. 29).

The distribution of sherds leads to the inference that there
was a residential zone* in and around Area 40a and possibly in Area
29b. The strip along the canal between these two Areas was probably
farmed by irrigation and fertilized with household trash. In the Upper
Sector, the zones around Areas 8c and 9 (which are Level 3) could have
been especially favored with household trash, but more likely were
residential areas for a much shorter time than was Area 40a. It is
also possible that Areas 8c and 9 were collected more intensively
than surrounding Areas due to small-Area bias. However, this kind of
bias was found not to be statistically significant (see chapter XI).

LATE MONTE ALBAN I PHASE

During the Late Monte Albán I phase the highest ceramic densities
occur, and the greatest surface area of the survey zone is occupied. The
same Areas which had sherds in Early Monte Albán I also had sherds in
Late Monte Albán I, but the density was much greater. Note the similar
distribution on the Early Monte Albán I Levels Reset Map (fig. 28) and
the Late Monte Albán I Primary Map (fig. 30).

In the Lower Sector, most of the Areas that were Level 2 (10 to 30)

_____

* Residential zone refers to land inferred to have been actually occupied by houses and cooking-midden areas. Surrounding these houses and
middens one would expect to find an area of associated debris. This
would be the zone with densities of Level 2 surrounding the residential
zone centered in Area 40a.

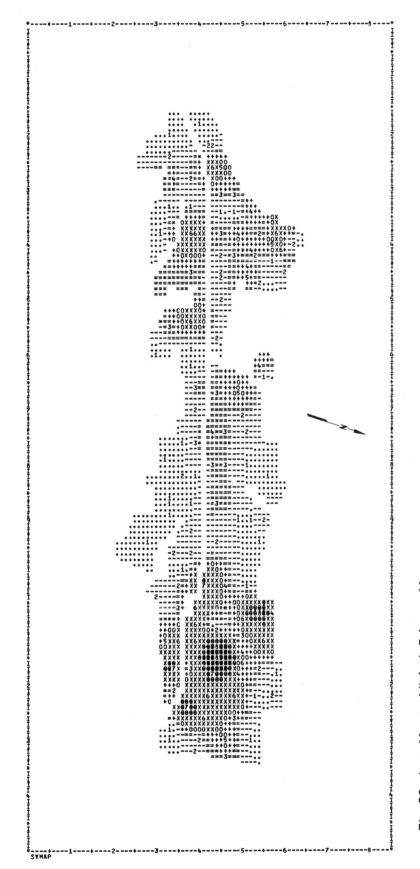

Fig. 30. Late Monte Albán I, Primary Map

or higher on the Early Monte Albán I Primary Map (fig. 27) are Level 6 (100 to 200) or higher on the Late Monte Albán I Primary Map (fig. 30). Within this cluster of Level 6 Areas, there are five nodes of Level 7 (200 to 300) or 8 (300 to 1000). The highest density is found in Area 40a, which has a density value of 673. This Area also had the highest density during Early Monte Albán I. Adjacent to Area 40a are Areas 39a and 41a, with density values of 361 and 264, respectively. The four other nodal Areas in the Lower Sector are Areas 46a, 53, 54a, and 37a, with density values of 345, 254, 235, and 220, respectively. As stated above, these Areas are mostly surrounded by Areas with Level 6 densities. However, Areas 38a, 38c, and 39b, which are Levels 2, 3, and 4, respectively, are exceptions. Area 38a, which has the lowest value of any Area on the ridgetop in the Lower Sector, is the location of Structure 7, a large mound.

The surface area of Level 2 or higher densities in the Lower Sector is 136,375 square meters. This is 84.1% of the total area of the Lower Sector (161,950 square meters). The Early Monte Albán I percentage for Level 2 or higher densities was 29.4%, and very little of that was above Level 2. The surface area which has Level 6 or higher densities is 46,625 square meters. This is 28.8% of the Lower Sector, contrasting with 1.9% in Early Monte Albán I. Areas with Level 7 or higher occupy 21,500 square meters, and Areas with Level 8 or higher densities occupy 6,200 square meters. These are 13.3% and 3.8%, respectively, of the Lower Sector surface area.

In the Middle Sector, the highest density occurs in Area 29b, which is Level 5, with a density value of 91. There is also a strip along the canal on the north side which is entirely Level 3. It comprises Areas 29a, 3a, 3b, 3c, 3d, and 36a. Area 3a is Level 4, however, with a value of 50. Consistently lesser values (Levels 1 and 2) occur further away from the canal to the north. Level 2 or greater densities occupy 101,850 square meters or 53.3% of the Middle Sector. Level 5 occupies 4,725 square meters. There is no Level 6.

In the Upper Sector, the Early Monte Albán I node in Area 9, which had been Level 3, increased to Level 5 (80 to 100), with a value of 92. Areas 13 and 15, which are adjacent to Area 9, are Level 6, with values of 133 and 139, respectively. Extending south from Areas 13 and 15 are Areas 10c, 11b, and 19c, which are all Level 4 (50 to 80). However, Area 12, in the middle of this Level 4 cluster, is Level 2, with a value of 28. Area 12 is the location of Structure 4. Structure 1 is just across the boundary from Area 12 in Area 11b. The other Level

6 Areas in the Upper Sector are Areas 2, 5a, 5b, and 21a. These were
all Level 2 in Early Monte Albán I. Area 5 has the highest densities
of any Area in the Upper Sector; Area 5a has a density value of 160,
and Area 5b has a density value of 180. The general pattern in the
Upper Sector is characterized by four nodes of Level 6 density which
are separated from each other by zones of Levels 1 and 2. Level 2 or
above densities occupy 122,815 square meters, which is 77.2% of the
total surface area of the Upper Sector (159,240 square meters). This
figure was 30.2% for Early Monte Albán I. Level 6 occupies 18,400
square meters or 11.6% of the Upper Sector.

The total surface area of all Areas collected is 512,615 square
meters. Level 2 or greater densities were found on 361,040 square
meters or 70.3% of the surface area. Level 5 or greater densities
occupied 85,750 square meters or 16.7%, and Level 6 or greater densities
occupied 65,025 square meters or 12.7% of the total surface area. This
latter figure compares with 0.6% for the Early Monte Albán I phase.

The Levels Reset Map (fig. 31) shows the decrease in values away
from the north side of the canal in the Middle Sector much more markedly
than the Primary Map. Going from Area 3c to Area 3d to Area 31, the
reset levels go from Level 6 (35 to 75) to Level 5 (20 to 35) to Level
2 (1 to 4). The same pattern shows up on the south side of the canal
to a lesser degree. Area 65a is Level 5 (20 to 35), Area 66a is Level
3 (4 to 10), and Area 67 is Level 2 (1 to 4). Note that on this map,
Level 1 is zero only, so that Areas where absolutely nothing was found
can be distinguished. In Early Monte Albán I, 55 of the 121 Areas
had a density value of zero, whereas in Late Monte Albán I there were
only 16.

The Corrected for Vegetation Map will be discussed by vegetation
types. The correction factor for grass caused significant increases in
density values in Areas 19c, 20a, 20b, 29a, 29b, 33a, 37b, 43a, and
43b. Areas 19c, 29b, and 43a all increased to Level 8 (300 to 1,000),
with values of 321, 411, and 781, respectively. Before correcting for
vegetation, Area 19c had a value of 71 (Level 4), Area 29b had a value
of 91 (Level 5), and Area 43a had a value of 173 (Level 6). Areas
20a, 20b, 29a, and 43b increased from Level 3 to Level 6. Area 37b
increased from Level 4 to Level 7, and Area 33a increased from Level 2
to Level 5.

The correction factor for bushes (mostly thornbushes) caused
significant increases in the density values for Areas 7, 22a, 22b, 24a,
24b, 25a, 25b, and 26, all located in the Upper Sector. Area 24b in-

166

Fig. 31. Late Monte Albán I, Levels Reset Map

167

Fig. 32. Late Monte Albán I, Corrected for Vegetation Map

SYMAP

creased from Level 4 to Level 8, Areas 7, 22a, 22b, and 24a increased from Level 2 to Level 6, Area 25a increased from Level 1 to Level 5, and Areas 25b and 26 increased from Level 1 to Level 3. However, as noted above, the thornbush Areas could not be collected systematically, and the parts which were collectible may have been overcollected. Thus, there is much room for skewing when the results are multiplied by such a large correction factor. Much less confidence should be placed in the corrected values for bushes than in those for grass, even though they are numerically similar.

The only other vegetation type which significantly increased density values was peanuts. Only seven Areas were covered by peanuts. Three of these (Areas 28b, 61, and 64b) had no Late Monte Albán I sherds. The other four all showed significant increases when vegetation was corrected for. Area 48a went from Level 2, with a density value of 28, to Level 8, with a value of 387. Area 48b went from Level 1, with a value of 5, to Level 4, with a value of 69. Area 48 is in the Lower Sector near other high-density Areas, so these vegetation corrections may have some validity. Area 66a went from Level 1, with a value of 6, to Level 5, with a value of 83. Area 66b went from Level 2, with a value of 10, to Level 6, with a value of 138. Area 66 is on the south side of the canal in the Middle Sector and is not adjacent to any other high-density Areas. Thus, it is probable that in this Area no higher densities are actually concealed by the vegetation.

LATE MONTE ALBAN I: DISCUSSION

The distribution of sherds in Late Monte Albán I is essentially the same as in Early Monte Albán I, but densities are much higher. The orientation with respect to the canal in the Middle Sector continues, as does the high-density node in Area 40a. Area 40a went from a density value of 102 in Early Monte Albán I to a value of 673 in Late Monte Albán I. Both of these density values are the highest for their respective phases. Area 40a is surrounded by other "satellite nodes," all in a large zone with density values above 100 (Level 6 and greater). There are four Level 6 high-density zones in the Upper Sector, separated from each other by much lower values. There may also be a high-density zone at the upper or west end of the Middle Sector in Area 29b hidden by grass (see the Corrected for Vegetation Map, fig. 32).

There are some interesting associations between sherd distribution and surface features such as mounds and terrace walls (see fig. 4 for

location of mounds and terrace walls). All of the mounds in the survey
area are in low-density Areas and are adjacent to high-density Areas.
Structures 2, 3, and 4 are in Area 4a, which is Level 3 and is adjacent
to Area 2, which is Level 6. Structure 7 is in Area 38a, which is Level
2 and is immediately adjacent to Areas 39a and 40a, which are Level 8.
Structure 8 is in Area 12, which is Level 2 and is next to Area 13,
which is Level 6. Structures 9 and 10 are in Area 1, which is Level
3. Adjacent Areas are Area 5 to the west and Area 2 to the east; both
of these Areas are Level 6.

Terrace walls are related to medium- and high-density Areas in the
Middle and Lower Sectors, whereas in the Upper Sector they are ubiquitous.
The Upper Sector has a steeper slope, so that there is a greater need
for terracing. In the Middle Sector, a terrace wall separates Area 29
from Areas 3 and 30. There is a relatively high-density zone in Area
29b (which may have been as high as Level 8 if vegetation is taken into
account), while Area 3 is part of the postulated agricultural zone
along the canal. The terrace wall could have separated agricultural
and residential zones. There is also a terrace wall running parallel
to the canal separating Areas 3 and 36 from Areas 30, 31, 32, and 35,
which are lower and slope toward the barranca. Areas 3 and 36 are
relatively level Areas adjacent to the canal. The terrace wall
probably functioned to retain soil and water near the canal along the
ridgetop in Areas 3 and 36. These two Areas are the Areas where Level
3 densities are found forming a strip along the canal, contrasting
with lower Level 1 and Level 2 densities on the other side of the
terrace wall in Areas 30, 31, 32, and 35. As in Early Monte Albán I,
this distribution is taken to indicate the dumping of household rubbish
on intensively used agricultural land along the canal. The association
of canal, terrace wall, topography, and sherd distribution is consistent with this interpretation.

In the Lower Sector, the only terrace walls are those along the
eastern edge of Area 37 and the parallel set of walls separating Areas
38 through 41 from Areas 51 and 52. This set of walls protects
Structure 7 and the high-density Areas 39a, 40a, and 41a from runoff
from further up the ridge and channels it past these Areas between
the parallel walls. Areas 39a, 40a, and 41a occupy the ridgetop, while
the canal runs below it to the south. Thus, both the canal and the
terrace walls are placed so as to keep water away from these Areas. This,
along with the high sherd densities, leads one to categorize this zone
as residential.

It should be pointed out, however, that the terrace walls are not dated, but their locations, as discussed above, best fit the Late Monte Albán I sherd distribution. Most piedmont canal-irrigated fields in use today are not terraced (Kirkby 1973), so it is probable that terrace walls would be built only during the most-intensive use of the area. Assuming this correlates with the greatest occupational density in the area, this would have been during Late Monte Albán I. Chronological placement of the mounds is not quite so inferential, since surface collections were made from each of them. These surface collections indicate a Late Monte Albán I date for the construction and use of all mounds except Structures 3, 5, and 6. Structure 3 had a preponderance of Monte Albán II sherds, with some Late Monte Albán I sherds. It may have been built during Late Monte Albán I and continued in use during Monte Albán II. Structures 5 and 6 (not discussed above) probably were constructed during Monte Albán V.

In order to further discuss residential areas, it is necessary to review the Early Monte Albán I pattern. The highest density in Early Monte Albán I was Level 6 in Area 40a, and it was suggested that if any Area was residential, it must have been Area 40a. Supporting this classification of Area 40a as residential in Early Monte Albán I is the clear break in density values. The Early Monte Albán I value in Area 40a was 102, and the next highest value elsewhere was 31. This indicates the possibility of a dichotomy between residential and non-residential density values. If this is the case, a similar dichotomy should be discernible in Late Monte Albán I. Table 32, which gives the frequency distribution of collection units by level for Late Monte Albán I (on the Primary Map), illustrates that the number of collection units per density level decreases as the density of the levels increases, until Level 5, with only six collection units falling within it, is reached. The next level, Level 6, has 16 collection units. This could be taken to indicate a break in density values at Level 5, so that Level 6 and greater densities (greater than 100) could be residential. This is not realistic, however, since the density levels do not encompass equal ranges of values.

To see if a break in density values still occurred when the ranges of values were equivalent, a frequency distribution of Late Monte Albán I density values by groups of 20 was made (table 37). The number of cases assignable to each group of 20 decreases as the values increase, until between values of 60 and 80 only three cases occur. Then, in the next group (80 to 100), six cases occur and, in the following two,

Table 37. Frequency distribution of density values by groups of 20
for Late Monte Albán I

Density Value Groups	Number of Cases (Units)
0 - 19	55
20 - 39	22
40 - 59	14
60 - 79	3
80 - 99	6
100 - 119	5
120 - 139	5
140 - 159	0
160 - 179	3
180 - 199	3
200 - 219	0
220 - 239	2
240 - 259	1
260 - 279	1
280 - 299	0
300 - 319	0
320 - 339	0
340 - 359	1
360 - 379	1
673	1

five each. Then there is another break between 140 and 160 with zero
cases, followed by groups with zero, one, two, or three cases up to
a value of 380. Finally, there is one case with a value of 673. A
break in the continuum of density values occurs between 60 and 80 and
between 140 and 160. This indicates that Areas with density values of
Level 5 and greater (values above 80) or the upper half of Level 6 and
above (values above 150) may be residential. The assumption is that
there is a density which is characteristic of residential use of an
Area and a density which is characteristic of a scattering of sherds in
agricultural areas. There should be relatively few Areas which fall
in neither category. During Late Monte Albán I, the density values
do distribute into a high- and a low-density group, with the break
between the two groups occurring around Level 5 (values of 80 to 100).
The spatial distribution of the values supports the statistical distribution. During Late Monte Albán I, Level 6 or greater (values
above 100) zones are nodal in the sense that they are surrounded by
Areas of much lower density, usually Level 1 or 2 (fig. 33). Thus,
both statistically and spatially, there seems to be a break in the

Fig. 33. Late Monte Albán I, map of residential Areas defined as having density values greater than 100

distribution of density values at Level 5, which may be indicative of a dichotomy in density values between residential and nonresidential zones. The other break in the distribution of density values may be due to the amount of time an Area was occupied.

If Areas with densities of Level 6 and greater are taken to be residential, then 6.5 hectares or 12.7% of the total surface area was residential (fig. 33). If Level 5 and greater densities are taken to be residential, then 8.6 hectares or 16.7% was residential. Level 5 includes Area 29b in the Middle Sector, which, when vegetation is allowed for, could have a density value as high as 411 (Level 8). The other corrections for vegetation have little effect on the density values outside of Area 19c and the thornbush Areas (Areas 7, 22, 24, 25, and 26). As discussed above, the Corrections for Vegetation represent the maximum possible error, and the corrections for the thornbush Areas are much less probable than the corrections for the other vegetation types.

The general pattern leads to the inference that the probable Early Monte Albán I residential zone in Area 40a greatly increased in size, and that substantial expansion to the Upper Sector occurred in Late Monte Albán I. Mounds are found in low-density Areas directly adjacent to high-density Areas which are probably residential. These structures may have had a ceremonial function but were probably high-status residences. The lower densities in the Middle Sector make it likely that this sector was mostly reserved for agriculture.

MONTE ALBAN II PHASE

The Primary Map for Monte Albán II (fig. 34) shows that most of the same Areas were occupied in Monte Albán II as in Late Monte Albán I, but at greatly reduced densities. In the Lower Sector the Late Monte Albán I high-density nodes (Level 8) were Areas 40a, 46a, 53, and 54a. During Monte Albán II, Areas 46a and 54a are Level 4, with density values of 55 and 51, respectively. Area 53 is Level 3, and Area 40a, which had the highest density value in both Early and Late Monte Albán I, is only Level 2 during Monte Albán II. Area 42b, which was Level 6 in Late Monte Albán I, is Level 3 during Monte Albán II. The rest of the Lower Sector has density values falling in Levels 1 and 2.

The Upper Sector Late Monte Albán I Level 6 nodes in Areas 2, 5b, 15, and 21 continued to have the highest densities in the Upper

SYMAP

Fig. 34. Monte Albán II, Primary Map

175

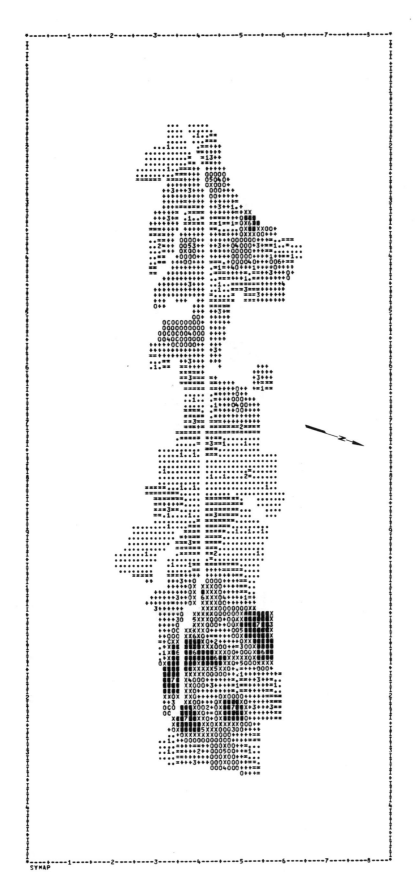

Fig. 35. Monte Albán II, Levels Reset Map

176

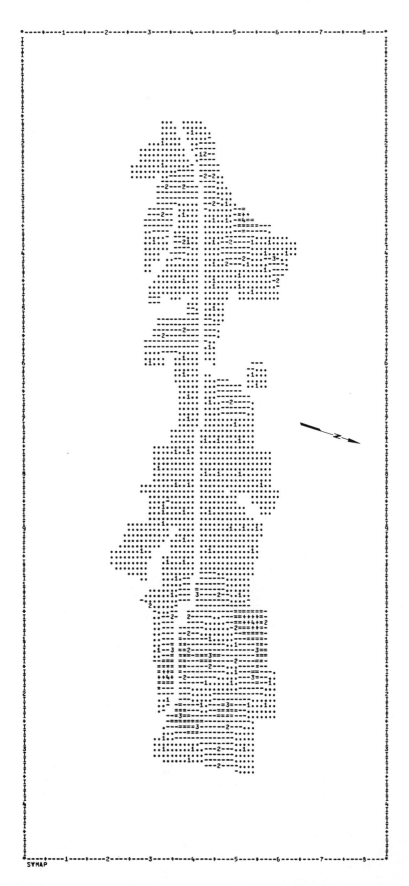

Fig. 36. Monte Albán II, Corrected for Vegetation Map

Sector in Monte Albán II. However, the highest densities are only Level 2 (density values between 10 and 30). The highest density values in the Upper Sector are 27 in Area 19c and 26 in Area 15. Areas 4a, 10b, 10c, and 11b are also Level 2. All other Areas in the Upper Sector have density values falling in Level 1. The Middle Sector is all Level 1 except Area 29b, which is in the lower range of Level 2, with a density value of 13. This Area also had the highest density value in the Middle Sector during Late Monte Albán I.

Areas with densities in Levels 3 and 4 are found only in the Lower Sector. Level 4 occupies 4,000 square meters, and Level 3 occupies 6,750 square meters. Level 3 and greater densities occupy 2.1% of the total area surveyed. The surface area of Areas with Level 2 densities is 84,525 square meters. Level 2 and greater densities occupy 95,275 square meters, comprising 18.6% of the total area surveyed.

Zoning along the canal (higher density values nearer the canal than farther away from it) disappears in Monte Albán II, although a highly attenuated form of it may be represented on the Levels Reset Map (fig. 35). This map also shows the distribution of sherds in the Lower Sector more clearly. Areas 42b, 46a, 53, and 54a show up as maximums (Level 7) on the peripheries of the Lower Sector, with lower values in the center.

The Corrected for Vegetation Map (fig. 36) shows little change from the Primary Map (fig. 34). The only Area which increased more than one density level is Area 19c (a grassy Area), which went from Level 2 to Level 4.

MONTE ALBAN II: DISCUSSION

The fact that the same Areas which had sherds during Late Monte Albán I also had them in Monte Albán II, and that they were distributed in approximately the same proportions among the Areas, shows that there was not a settlement shift between the two phases. However, the greatly reduced densities require discussion. Either much fewer people were doing the same thing as during Late Monte Albán I throughout Monte Albán II, or approximately the same number of people as in Late Monte Albán I were there for a very short time at the beginning of Monte Albán II. The latter seems more likely, since if even a few people were living in the same place throughout the entire 400-year duration of Monte Albán II, one would expect at least a Level 6 node to appear, since Early Monte Albán I produced a Level 6 Area and lasted only about 150 years.

There is evidence from the canal cuts that the canal was abandoned at the end of Late Monte Albán I or the beginning of Monte Albán II. Sherds from the canal fill were mostly Late Monte Albán I, plus a few Monte Albán II sherds. The surface collections corroborated this, since the distribution of sherds shows much less relationship to the canal than in the previous two phases.

MONTE ALBAN III-A PHASE

Only 51 Monte Albán III-A rim sherds were collected from only 23 of the 121 Areas. The only Areas that show up as Level 2 or greater on the Primary Map (fig. 37) are Areas 8c, 8d, 8e, 9, 15, and 39a. All of these Areas are located in the Upper Sector except Area 39a, which has the highest density value (34), which is Level 3. Area 8e is also Level 3, with a density value of 33, and the rest of the Areas mentioned above are Level 2. The Middle Sector is entirely lacking in Monte Albán III-A sherds except in Area 3a, which has a density value of 3. The surface area of the Areas with Level 3 densities is 3,600 square meters. Level 2 densities occupied 8,500 square meters and Level 2 and Level 3 densities together occupied 12,100 square meters or 2.4% of the total surface area.

The Corrected for Vegetation Map (fig. 38) shows five Areas raised to Level 6. These are Areas 15, 27a, 27b, 39a, and 43a. Areas 27 and 43a are covered by grass, Area 15 is covered by low corn, and Area 39a is covered by high corn.

MONTE ALBAN III-A: DISCUSSION

The survey area seems to have been largely abandoned for settlement at this time. There is little evidence for spreading of household rubbish, and the higher-density Areas which might be called residential are only Level 3. The corrections for vegetation produce some Level 6 Areas, but it must be remembered that the Corrected for Vegetation Maps show the maximum possible error which could have been caused by vegetation. There are two other factors which may also have biased the density values. Low densities during Monte Albán III-A may be the result of the fact that only a few ceramic types are known for Monte Albán III-A. When the ceramic sequence becomes more refined,

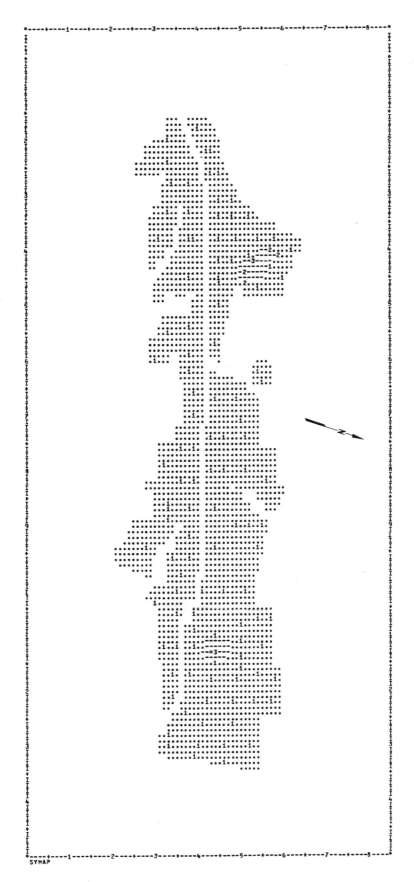

Fig. 37. Monte Albán III-A, Primary Map

180

Fig. 38. Monte Albán III-A, Corrected for Vegetation Map

some of the unidentifiable sherds may turn out to be Monte Albán III-A. On the other hand, some of the higher densities may be due to an increased-level-of-interest bias in Areas having overall high densities (when all phases are taken into account). See chapter XI for a discussion of this. When all of the above is taken into account, it may be inferred that during Monte Albán III-A there were a few scattered houses occupied for short periods of time within the 300-year duration of this phase.

MONTE ALBAN III-B/IV PHASE

Only 32 of the 121 Areas had Monte Albán III-B/IV sherds on them, and only 102 rim sherds were collected. The Primary Map shows that in the Lower Sector, Areas 39a and 40a are Level 3, with density values of 45 and 35, respectively. Areas 56, 57a, and 58 are Level 2. In the Upper Sector, Area 15 is Level 3, with a value of 35, and Areas 9, 10b, and 24a are Level 2. The Middle Sector has Monte Albán III-B/IV sherds in only three Areas, and these are all Level 1. The surface area of Areas with Level 3 is 6,075 square meters. The surface area of Areas having Level 2 densities is 21,075 square meters, and Level 2 or greater densities occupy 27,150 square meters or 5.3% of the total survey area. The corrections for vegetation had little effect other than increasing Area 56, which is covered by grass, from Level 2 to Level 4.

MONTE ALBAN III-B/IV: DISCUSSION

The Monte Albán III-B/IV pattern is almost the same as that of Monte Albán III-A, although a few more Areas have sherds and there are slightly higher densities in the Lower Sector. The Level 3 high-density zone in Area 39a in the Lower Sector continued and spread to adjacent Area 40a, and the Level 3 high-density zone in the Upper Sector moved from Area 8e to Area 15, adjacent to it. It was noted for Monte Albán III-A that the densities might actually be higher because of the few ceramic types assignable to Monte Albán III-A and the effects of vegetation in hiding sherds in several Areas. However, in Monte Albán III-B/IV, neither of these factors is a problem, so that the low densities are probably quite accurate. As in Monte Albán III-A, a sample-size bias may be operating (see chapter XI). Since the same pattern exists as in Monte Albán III-A, the same inference is indicated:

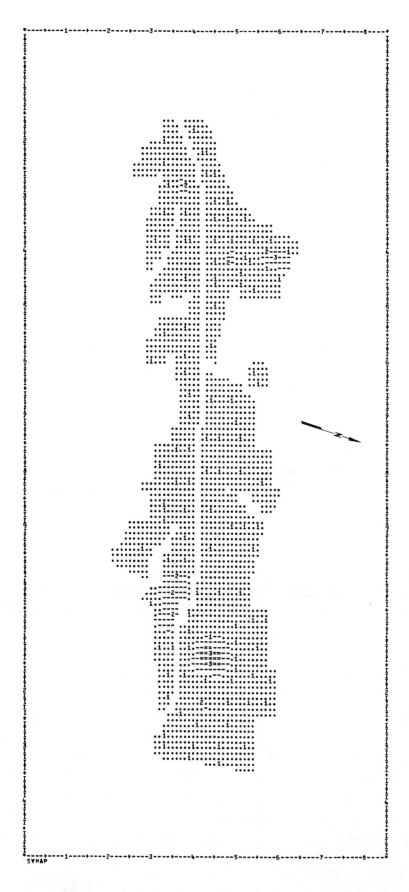

Fig. 39. Monte Albán III-B/IV, Primary Map

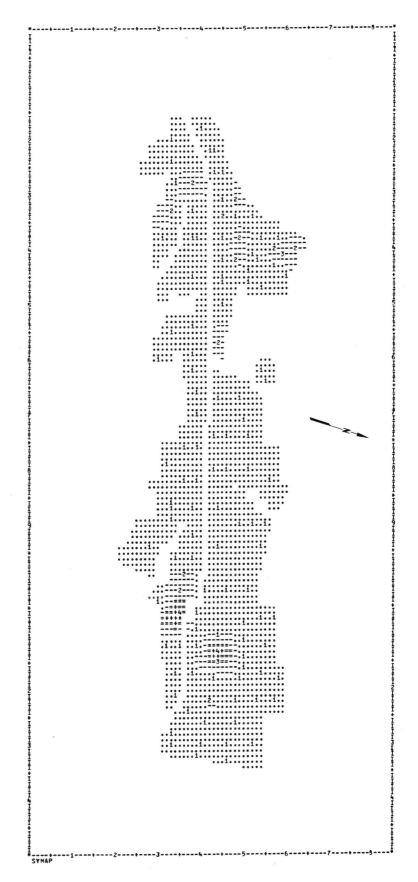

Fig. 40. Monte Albán III-B/IV, Corrected for Vegetation Map

scattered houses occupied for short periods of time.

MONTE ALBAN V PHASE

This phase is the only one during which the ridgetop around Areas 39a and 40a was not occupied (fig. 41). Areas 38 through 40 and Area 42 have density values of zero, and Areas 41a and 41b have values of four and three. The highest value occurs in Area 53, which is Level 5, with a density value of 85, and the second-highest density value is 55 (Level 4) in Area 46a. Areas 44b and 51b are Level 3. These are all single-node Areas of higher density on the peripheries of the Lower Sector, separated from each other by zones of Levels 1 and 2. In the Upper Sector, Areas 9 and 13 are Level 3.

There are 21 Areas which are Level 2 and 34 which are Level 1 (one or more sherds). The total number of Areas which have sherds is 61. Half of the 121 Areas have at least one Monte Albán V sherd, but 53 of them are only Level 1 or 2. The general pattern is a low-density random scatter with avoidance of the central part of the Lower Sector and the southern part of the Upper Sector. Only two Areas are really nodal, with density values in Levels 4 and 5. Level 5 occupies 1,775 square meters, and Level 4 occupies 1,275 square meters. Level 3 occupies 7,950 square meters, and Level 2 accounts for 70,100 square meters. Level 2 and greater densities are found over a surface area of 79,825 square meters or 15.6% of the total area surveyed.

The Corrected for Vegetation Map (fig. 42) shows that the corrections for vegetation had little effect on the values for this phase. The only Area which changed more than one level is Area 33a, which went from Level 2 to Level 4. Thus, there are no Areas where significantly higher densities may have been masked by vegetation.

MONTE ALBAN V: DISCUSSION

The low-density scatter of sherds with a few higher-density Areas indicates dispersed settlement similar to the pattern of dispersed single households characteristic of other parts of the Valley at this time (Dudley Varner, personal communication). Area 53 (Level 5) and Area 46a (Level 4) may represent single-family households which may have been engaged in rainfall agriculture. Increases in corncob size probably made rainfall agriculture in this area possible by this time.

Fig. 41. Monte Albán V, Primary Map.

SYMAP

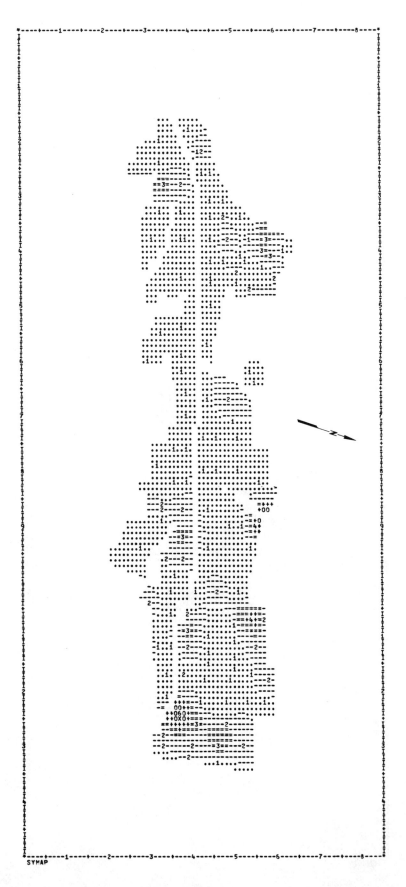

Fig. 42. Monte Albán V, Corrected for Vegetation Map

The avoidance of the center of the Lower Sector around Areas 39a and 40a is notable, since in every other phase this was the locus of highest density. It may be that the previous activity in this zone made it good agricultural land.

13. Estimated Carrying Capacity

To formulate estimates of the carrying capacity of the Xoxocotlán
piedmont, a number of factors had to be considered: (1) the amount of
land actually brought under irrigation-based cultivation; (2) the amount
of crops which could have been grown in this area; and (3) the average
amount of food eaten by one person in one year.

Estimates based on ethnographic cases are available for the maize
yield from irrigated land. In addition, the amount of maize consumed
by an average family (consisting of five persons) has been documented.
Using these figures, a rough estimate of maximum carrying capacity of
the approximately 50 hectares of irrigable land in the Xoxocotlán
piedmont can be determined.

This estimate represents optimal carrying capacity today if the
area were irrigated. By correcting for lower yields due to smaller
corncob size in the past, an estimate can be extrapolated for the Late
Monte Albán I phase, when the area was most heavily occupied.

ESTIMATES OF MODERN YIELDS AND CONSUMPTION REQUIREMENTS

A number of workers have studied the relationship between size of
fields and minimal yields required to support modern populations in
highland Mesoamerica.

Palerm (1955) states that among the Highland Totonac, a family of
five persons requires 0.86 hectare of irrigated land to raise enough
maize for daily nutritional requirements. Using these figures, the
population density for the area works out to approximately 5.8 persons
per hectare. Sanders' (1970) density figures for the population served
by the Teotihuacán irrigation system on alluvial land in the sixteenth
century is seven persons per hectare. For the modern Valley of Oaxaca,

Kirkby (1973) estimates the average maize yield for piedmont irrigation
systems to be 1.5 metric tons per hectare and the consumption requirements of a family of five to be 2.4 metric tons per year, if all subsistence requirements are figured in terms of maize alone. This is
3.1 persons per hectare.

Granskog (1974) provides much more detailed data from the village
of Santo Tomás Mazaltepec in the Etla arm of the Valley of Oaxaca.
The predominant form of agriculture there is piedmont irrigation, which
yields two crops per year. If alfalfa is not included in the cycle,
an almud of land (one-fourth hectare) is planted for four years and
allowed to lie fallow for three years. The first year produces a
winter crop of 25 canastas (baskets) per almud and a summer crop of
35 canastas, a total of 60 canastas per almud. The total yield the
second year is 50 canastas per almud, while the third year's yield is
40 and the fourth year's yield is 30. This is a total of 180 canastas
or 4,104 kilograms (one canasta equals 22.8 kilograms) from one-fourth
hectare over seven years (including the fallow period). Multiplying
by four and then dividing by seven gives 2,345 kilograms or about 2.35
metric tons per hectare per year (see table 38a).

Table 38. Yields from one-quarter hectare for (A) seven-year cycle and (B) nine-year cycle, for
piedmont irrigation of Santo Tomás Mazaltepec (according to Granskog 1974)

| | | A | | | | | B | | |
Year	No. of Crops	Yield in Canastas	Yield in Kgs.		Year	No. of Crops	Yield in Canastas	Yield in Kgs.
1	2	60	1368		1	2	60	1368
2	2	50	1140		2	1	30	684
3	2	40	912		3	2	50	1140
4	2	30	684		4	1	30	684
5	0	0	0		5	1	20	456
6	0	0	0		6	0	0	0
7	0	0	0		7	0	0	0
		---	----		8	0	0	0
		180	4104		9	0	0	0
							---	----
							190	4332

An alternate cropping method involves the planting of two crops
the first year (yield: 60 canastas per almud), one crop the second
year (yield: 30 canastas), two crops the third year (yield: 50 canastas), one crop the fourth year (yield: 30 canastas), and one crop

the fifth year (yield: 20 <u>canastas</u>), for a total yield of 190 <u>canastas</u>
(4,332 kilograms) per <u>almud</u> for five years. The field is then left
fallow for four years, making a nine-year cycle. Multiplying by four
and then dividing by nine gives 1,925 kilograms or about 1.9 metric
tons per hectare per year (see table 38b).

Brennan (1971) also studied yields in Mazaltepec and provides data
on yields for a seven-year cycle which includes two years of fallow.
This approximates the Prehispanic yields, according to Brennan, since
manuring was not practiced. If it were, irrigated land would not
require a fallow period. Brennan presents his data in terms of
<u>carretas</u> of ears of corn (still in the husk). One <u>carreta</u> is equal
to about 600 kilograms of corn kernels. One-fourth hectare produces
1.2 <u>carretas</u> the first year, 0.9 <u>carretas</u> the second, third, and
fourth years, and 0.6 <u>carretas</u> the fifth year. This is a total of
1.54 metric tons per hectare per year over the seven-year cycle (see
table 39).

Table 39. Yields from one-quarter hectare for a seven-year cycle for
piedmont irrigation of Santo Tomás Mazaltepec (according
to Brennan 1971)

Year	Yield in Carretas	Yield in Kgs.
1	1.2	720
2	0.9	540
3	0.9	540
4	0.9	540
5	0.6	360
6	0.0	0
7	0.0	0
	4.5	2700

Kirkby (1973) states that the average family of five eats about
one metric ton of corn per year, but that this is only 50% (monetarily)
of total annual food needs, since meat and vegetables are also consumed.
For reconstruction of past consumption requirements figured entirely
in corn, Kirkby suggests 2.4 metric tons per family of five per year.

Granskog (1974) states that a family of five eats about 40
tortillas a day. Forty tortillas can be made from one <u>almud</u> (one-sixth
of a <u>canasta</u>) of corn (there are two meanings for the word <u>almud</u>). This
represents about 3.8 kilograms per day or 1,380 kilograms per year (1.38
metric tons). If extra corn must be grown to exchange for other foods,
as Kirkby suggests, then perhaps 1.5 <u>almudes</u> or one-fourth of a <u>canasta</u>

per day would be necessary. At this rate, 2.08 metric tons would be required per year. Brennan (1971) estimates a family's annual consumption of corn to be 1.83 metric tons.

In summary, piedmont irrigation in the Valley of Oaxaca has been estimated to annually produce 1.5 metric tons per hectare (Kirkby), 1.54 metric tons per hectare (Brennan), 1.92 metric tons per hectare (Granskog, using the nine-year cropping cycle), and 2.35 metric tons per hectare (Granskog, using the seven-year cropping cycle). Consumption of corn has been estimated to be 1.38 metric tons per year (Granskog), 1.83 metric tons per year (Brennan), 2.08 metric tons per year (Granskog), and 2.4 metric tons per year (Kirkby) for a family of five persons. If the low, middle, and high estimates for production and consumption are paired, it will be seen that the carrying capacity is, in each case, about one family of five per hectare.

The above discussion has shown that the average annual yield of a seven-year cycle will support a family of five on one hectare of land. However, if a family of five owned only one hectare, could the fallow periods be arranged so that enough corn could be grown every year? Table 40 shows the three possible arrangements for the seven-year cycle with four quarter-hectare plots in use. It can be seen that the second arrangement yields 80 or more canastas every year (1,824 kilograms), while the other arrangements have years with 70 canastas (1,596 kilograms), 60 canastas (1,368 kilograms), and 30 canastas (684 kilograms). If consumption requirements for a family of five are taken to be 2,000 kilograms per year, a family of five could probably exist on one hectare using the second arrangement, but would fail to grow sufficient corn for subsistence in some years using either of the two other arrangements. Where irrigable land is scarce, there would, therefore, be a definite advantage in planning and coordinating planting and fallow cycles of the plots of land held by a family, and throughout the community as well. Today, units of production are independent households, and the function of the community government is usually restricted to the settling of disputes. However, in Prehispanic times, communitywide coordination of planting may have been possible.

Table 40. Annual yields for three possible arrangements of fallow cycles for four quarter-hectare plots

| | YIELD IN CANASTAS | | | | | Yield | Consumption |
Year	Plot 1	Plot 2	Plot 3	Plot 4	Total	in Kgs.	Difference*
1	60	0	0	0	60	1368	- 632
2	50	60	0	0	110	2508	+ 508
3	40	50	60	0	150	3420	+1420
4	30	40	50	60	180	4104	+2104
5	0	30	40	50	120	2736	+ 736
6	0	0	30	40	70	1596	- 404
7	0	0	0	30	30	684	-1316
1	60	0	30	50	140	3192	+1192
2	50	0	0	40	90	2052	+ 52
3	40	60	0	30	130	2964	+ 964
4	30	50	0	0	80	1824	- 176
5	0	40	60	0	100	2280	+ 280
6	0	30	50	0	80	1824	- 176
7	0	0	40	60	100	2280	+ 280
1	60	0	50	0	110	2508	+ 508
2	50	0	40	0	90	2052	+ 52
3	40	0	30	60	130	2964	+ 964
4	30	60	0	50	140	3192	+1192
5	0	50	0	40	90	2052	+ 52
6	0	40	0	30	70	1596	- 404
7	0	30	60	0	90	2052	+ 52

* Difference from consumption requirement of 2000 kgs.

Table 41. Carrying capacities for yields between 500 and 2000 kgs. per year (assuming a consumption requirement of 2000 kgs. per year for a family of five)

Yield in Kgs.	No. of Ha Required by Family of Five	Persons/Ha.
500	4.00	1.25
600	3.33	1.50
700	2.85	1.75
800	2.50	2.00
900	2.22	2.25
1000	2.00	2.50
1100	1.81	2.75
1200	1.67	3.00
1300	1.53	3.25
1400	1.42	3.50
1500	1.33	3.75
1600	1.25	4.00
1700	1.17	4.25
1800	1.11	4.50
1900	1.05	4.75
2000	1.00	5.00

194

ESTIMATED CARRYING CAPACITY OF THE XOXOCOTLÁN
PIEDMONT DURING THE LATE FORMATIVE PERIOD

Up to this point, all data and estimates have been made with
reference to present conditions. In order to discuss the possible
carrying capacity of the Xoxocotlán piedmont during the Late Monte
Albán I phase, certain assumptions and corrections must be made. If
the Mazaltepec data are to be applied to the Xoxocotlán piedmont, it
must be assumed that soil and water conditions are similar, especially
in terms of nutrient content. Lacking soil and water analyses, it
can only be stated that both areas are in the same physiographic zone.
It must further be assumed that these conditions were the same between
550 and 150 B.C. as they are now. Pollen data show that the climate
of the Valley of Oaxaca has been very similar to that of the present
climatic regime since at least 1100 B.C., although the period from
400 to 1 B.C. was slightly wetter (Flannery and Schoenwetter 1970).
Another possible difference to be considered is the amount of water
available for irrigation. Mazaltepec is on a perennial stream which
provides water during the dry season, thus making two crops a year
possible. The Xoxocotlán system did not have a perennial stream as
a water source, but depended on runoff impounded behind a dam. Thus,
it must be assumed that the dam impounded enough water to make two
crops a year possible on all land irrigable from the canal and which
was not fallow, in order to make use of the figures presented above,
which assume two crops per year.

Finally, corncob size must be corrected for. Modern cobs range
up to 14 centimeters in length and yield up to three metric tons per
hectare per year. However, during the Late Monte Albán I phase, cob
size was only eight centimeters, as reconstructed by Kirkby (1973:
fig. 48), who presents a graph of cob length over time (based on
corncobs found in datable archaeological contexts). Kirkby's figure
48 also shows yields obtainable from various cob sizes. Kirkby states
that she was able to find these various shorter cobs growing today in
the Valley of Oaxaca and was thereby able to determine yields for them.
However, she presents no information about what type of agriculture
produced the yields for the various cob sizes. Thus, the graph shows
that eight-centimeter cobs yield 500 kilograms per year, but we are not
told whether this is from dry farming, high-water-table agriculture,
piedmont irrigation, or alluvial irrigation, nor whether this yield is
from one or two crops per year. As Kirkby points out at length else-

where in her monograph, modern yields vary considerably, depending on
availability of water. One must assume that this was also true in the
past. Thus, even though maximum cob length was eight centimeters in
500 B.C., yields could be increased by increasing water and nutrient
availability, as with irrigation, so that more cobs per hectare
could be produced. It seems reasonable to assume that the small cobs
Kirkby found were growing only in marginal rainfall-based agricultural
areas, since smaller cobs probably require less water or fewer nutrients
in the soil. If this were not the case, there would be no reason for
planting the smaller-sized corn. Therefore, it will be assumed here
that eight-centimeter cobs produce 500 kilograms of corn from one
rainfall-dependent crop per year. Yields from piedmont irrigation
during Late Monte Albán I times, with two crops per year, must be
somewhere between 500 kilograms and the modern yield averaging 2,000
kilograms. Table 41 shows the carrying capacity in persons per
hectare for yields between 500 and 2,000 kilograms per hectare per
year, assuming a consumption requirement of 2,000 kilograms per family
of five per year. Thus, the carrying capacity of the Xoxocotlán piedmont in Late Formative times was between 1.25 and five persons per
hectare.

14. Summary and Conclusions

The purpose of this chapter is to integrate the survey and excavation data and to place this information within the context of developments elsewhere in the Valley of Oaxaca. The chapter will conclude
with a discussion of the role of irrigation in the development of
Monte Albán as a major urban center. Fieldwork and subsequent
analyses indicate a community along an irrigation canal occupied during
Early and Late Monte Albán I phases, with areas of high ceramic density,
mounds, and terrace-wall constructions.

The Early and Middle Formative periods* (1400-550 B.C.) in the
Valley of Oaxaca can be seen as that period of time when agriculturally
based villages were becoming the dominant form of settlement. These
early villages are sparsely represented in the Tierras Largas phase,
where four or five are known for the Etla arm, but by the end of the
Guadalupe phase (700 B.C.) there are 14 sites in the Etla arm (Dudley
Varner 1974). These early villages are restricted to the low and high
alluvium along the Atoyac River, with agriculture based on floodwater
diversion and the high water table, which allows pot irrigation (see
Kirkby 1973). Pot irrigation has been demonstrated to be at least as
old as the Guadalupe phase (850-700 B.C.) by Orlandini (1967), who found
the outline of an ancient well preserved in an arroyo bank. The well
contained Guadalupe phase sherds.

Most of the Early and Middle Formative villages cover only two to
three hectares and are composed of an average of 10 households (Winter

---

* Represented in the Valley of Oaxaca by the Tierras Largas (1400-1150
B.C.) and San José (1150-850 B.C.) phases, and the Guadalupe (850-700
B.C.) and Rosario (700-550 B.C.) phases, respectively.

1976:234). Burial and residential data indicate little status differentiation within these villages (Winter 1974a). There was, however, a regional center at San José Mogoté in the Etla arm which was at least 10 times larger than the other villages, and had white-plastered hard clay and puddled adobe platforms and a magnetite working area with some status differentiation indicated (Winter 1974a; Flannery 1976b).

The Late Formative period shows a great increase in the number of sites (from 14 in the Guadalupe phase to over 40 in Early Monte Albán I), as well as the use of a new environmental zone (expansion into the piedmont zone). Status differences are indicated by variations in residences and burials (Winter 1974a). Expansion into the piedmont zone has been interpreted as marking the beginning of piedmont canal irrigation, making use of water in the perennial streams (Flannery and Schoenwetter 1970, Lees 1973). A new kind of site which appears during the Late Formative is the hilltop site with stone civic architecture. Many of these sites are strategically located overlooking the piedmont streams, leading to the inference that they were administrative centers located so as to control the water source for piedmont irrigation. However, no actual irrigation systems have been found in the Valley, except at Monte Albán (the Xoxocotlán system), and this system does not have a piedmont stream as a water source. The association in time between increased status differentiation, the appearance of administrative centers, and canal irrigation leads to the question of whether irrigation had a causal role in the development of this increased complexity.

Monte Albán represents a Late Formative hilltop administrative center with stone civic architecture, and the Monte Albán Survey (Blanton 1978) has shown it to have been a large residential center as well. It has an associated irrigation system (the Xoxocotlán system), but its water source was an artificial reservoir created by a dam. Thus, in the case of Monte Albán, it appears that settlement of the hilltop site was not undertaken in order to control a natural water source, but that an artificial water source and an irrigation system were created so that agriculture could take place close to the administrative-ceremonial center. The scale of the irrigation system also makes this the most likely interpretation. This will be discussed later in the chapter.

EARLY MONTE ALBAN I

Monte Albán was probably colonized by households from the Valley
floor, since there is a continuity in the organization and content
of household clusters between the Early and Middle Formative village
of Tierras Largas (located below Monte Albán to the north) and Early
Monte Albán I household clusters excavated at Monte Albán (Winter 1974b).
It is also significant that Tierras Largas was abandoned at the beginning
of Early Monte Albán I (Winter 1972). The Monte Albán Survey indicated
that "relatively dense Early Monte Albán I pottery" covered 69 hectares
but was spatially discontinuous, consisting of three discrete occupational zones to the east, west, and south of the Main Plaza. A less-
dense peripheral area covered 255 hectares (Blanton 1978:35). Caso,
Bernal, and Acosta (1967:96) reported the existence of an Early Monte
Albán I stone platform 2.5 meters high discovered during excavations
under the North Platform of the Main Plaza. It rests directly on bedrock
and is described as a structure "con muro en talud" with a drain carved
into the bedrock (ibid.). Tomb 33 also dates from the Early Monte
Albán I phase (ibid.:145) and contained 37 ceramic objects (Winter 1974a:
table 1). Both the stone platform and the tomb contrast with the simple
household clusters which were excavated by Winter (1974a, 1974b) northwest of the North Platform, each of which consisted of a 4 x 6-meter
house foundation with storage pits, ovens, and burials within a 10-
meter radius. Burials here had only a few ceramic objects. Thus,
during Early Monte Albán I, Monte Albán was already quite a large
hilltop administrative-ceremonial center with a large population and
marked status differentiation.

The Xoxocotlán piedmont also was the locus of an Early Monte Albán
I settlement at Monte Albán. However, it differed from the others in
that it was scattered along an irrigation canal which ran down the top
of a piedmont ridge. The zone of highest occupational density is
located at the lower end of the canal in Area 40a, which had a rim-
sherd density of 102 per hectare. As discussed in chapter XII, it is
assumed that Area 40a was residential, since it had such a high density
compared to all other areas (the next highest density was 43). It is
then debatable whether collection units with lower density values
should also be considered to be residential. If Early Monte Albán
I settlement was restricted to Area 40a, it would result in a residential
area of only about 3,000 square meters or 0.3 hectare. This would seem
to be too small an area to provide the population necessary to manage a

200

two-kilometer-long canal system. It will be recalled that Winter
states that most Early and Middle Formative Valley of Oaxaca villages
were two to three hectares in area and were composed of about 10
households or 50 people (Winter 1976:234).

If the next-lower group of density values (Level 3, or density
values between 30 and 50) is assumed to represent a residential
occupation (they were perhaps occupied for a shorter period of time
than Area 40a), then Areas 39a and 41a adjacent to Area 40a in the
Lower Sector and Areas 8c and 9 in the Upper Sector can be added
(see chapter XII for locations of Upper, Lower, and Middle Sectors).
This gives an area of 11,600 square meters or about 1.2 hectare.
Corrections for vegetation (fig. 29) indicate that some collection
units may have contained a higher density of Early Monte Albán I
sherds, but vegetation hindered collection. It is clear that not all
such collection units would actually yield a higher density under
ideal conditions (no vegetation), so these corrections cannot be
applied without some measure of caution. Four collection units in
addition to Area 40a have density values above 100 when corrected for
vegetation. However, one of these, Area 25a, is a thornbush area,
which could not be collected systematically (see chapter XII). The
other three collection units (Areas 29b, 43a, and 54a) have a combined
area of 13,350 square meters or about 1.3 hectare, which enlarges the
possible residential zone around Area 40a in the Lower Sector and
creates a new zone at the upper end of the Middle Sector (Area 29b).
If the Level 3 areas and the Corrected for Vegetation areas are both
assumed to be residential, the total Early Monte Albán I residential
area, including Area 40a, is 24,950 square meters or about 2.5 hectares.

A point which should be made here is that whatever density level
one assumes to represent a residential occupation, the decision can
be made on the basis of a quantitative comparison of density values
rather than on someone's subjective visual assessment of sherd distribution in the field, which has been characteristic of most regional
surveys in Mesoamerica (see chapter II).

A residential area of between 1.2 and 2.5 hectares is consistent
with Winter's (1976:234) statement that most Middle Formative villages
covered about two to three hectares (based on excavations). If the
Monte Albán area was colonized by such villages at the beginning of
the Late Formative period, it is probable that new agricultural
communities would be about the same size. The settlements on the top
of Monte Albán, which no doubt had some nonagricultural functions,

were much larger. Blanton (1978:35) estimates Early Monte Albán I occupation covered about three square kilometers, based on the subjective estimates of sherd distribution mentioned above.

Although the Early Monte Albán I Xoxocotlán settlement was about the same size as earlier villages, its arrangement was quite different. Instead of occupying a contiguous, roughly circular area (see Winter 1976: fig. 8.5), the Xoxocotlán settlement was dispersed along a ridgetop, with intervening nonresidential zones. This new pattern reflects a new agricultural technology: canal irrigation. The best agricultural land most likely is in the Middle Sector, since it is relatively level with deeper soil than further upslope. Conversely, much of the Lower Sector north of the canal could not have received water from the canal, since the canal is lower than the land surface in this zone. Therefore, the area was more suitable for residential than agricultural use. Distribution of sherds is consistent with these topographic features, as there is a light scatter of sherds along the canal in the Middle Sector, with more sherds closer to the canal than away from it, indicating an agricultural use of this area (with land closer to the canal being more intensively used, perhaps receiving more household trash as fertilizer). Area 40a, the most densely occupied area, is located in the portion of the Lower Sector which is north of the canal.

There are two lines of evidence which indicate that the canal was in use during Early Monte Albán I. Both of these are indirect, since evidence from excavation only dates its abandonment (sherds which washed into the canal after its abandonment; most were Late Monte Albán I, with a few Monte Albán II). The first line of evidence is the distribution of sherds discussed above, which seems directly related to the canal route. The second is that it is doubtful that the piedmont would have been cultivated at this time without the use of irrigation, due to the smaller corncob size, which would produce low yields in the poorer piedmont soil with only rainfall-dependent techniques. Kirkby's map of the predicted cultivated area during the Late Formative includes the piedmont zone only where she considered irrigation to be possible (along the perennial streams) (Kirkby 1973: fig. 51).

LATE MONTE ALBAN I

During Late Monte Albán I, Monte Albán became a major administrative-ceremonial-residential center covering over four square kilometers

(Blanton 1978:44). The discrete Early Monte Albán I communities coalesced around the Main Plaza area, which had taken on monumental proportions and was the focal point for construction of major civic-ceremonial buildings such as the Danzantes Building, with its massive stone construction and its associated bas-relief Danzantes and glyphs.

The Late Monte Albán I phase saw the largest occupation of the area around the canal on the Xoxocotlán piedmont. Most of the structures visible on the surface or encountered during excavation were constructed during this phase. Essentially the same pattern of occupation characteristic of Early Monte Albán I continued during Late Monte Albán I, but at a much denser level. Area 40a in the Lower Sector had the highest density during both Early and Late Monte Albán I, but during Late Monte Albán I its density was approximately seven times that of Early Monte Albán I. During Late Monte Albán I, a large zone around Area 40a had a density greater than 100 rim sherds per hectare, which was about 4.7 hectares in extent. Four smaller, less-dense zones with densities greater than 100 were located in the Upper Sector, while the Middle Sector was again the least-dense zone, probably still being reserved largely for agriculture. The total extent of zones with densities greater than 100 rim sherds per hectare is about 6.5 hectares.

As discussed in chapter XII, the locations of most mounds and terrace walls with respect to the Late Monte Albán I sherd-density distribution is consistent with a Late Monte Albán I date for their construction and use. Surface collections from the mounds (Structures 1-10) yielded mostly Late Monte Albán I sherds (table 1). However, Structure 3 contained a large percentage of Monte Albán II sherds, indicating continued use during that phase. Collections from Structures 5 and 6 indicate use of both structures during Monte Albán V, and it is possible that both structures were built during Monte Albán V. Excavation of Structures 9 and 9-sub-1 and the Area 1 terrace wall confirmed a Late Monte Albán I date for these structures. The presence of seven sherds transitional to Monte Albán II in the fill of Structure 9 indicates that it was constructed toward the end of Monte Albán I.

Terrace walls are common in the Upper Sector (where the slope is steepest) and run perpendicular to the canal route. In the Middle Sector, terrace walls are less frequent. The major terrace wall here runs parallel to the canal and probably served to keep irrigation

water and soil from running off the ridge into the north barranca.
Major terrace walls in the Lower Sector served to keep water away
from Structure 7 and the heavily occupied area surrounding Area 40a.
Since most piedmont irrigation systems in the Valley at the present
are not terraced (Kirkby 1973), the presence of terrace walls may
indicate that a slightly more intensive form of irrigation agriculture
was being practiced than at present.

The excavation of Structure 9 showed that it was a stone and
earth stepped platform covered with plaster. A hearth was found on
top of Structure 9 on the plaster floor, but it is possible that this
may date from a later period, as no associated sherds were found.
Structure 9 and the other Late Monte Albán I mounds (Structures 1, 2,
3, 4, 7, 8, and 10) may have had a ceremonial function, but were most
probably high-status residences, as were plastered platforms from
the Late Monte Albán I phase excavated at Huitzo and Tomaltepec
(Winter 1974a).

Most of these mounds are located in areas of low density which
are adjacent to areas of high density (over 100). This is especially
notable around Structure 7, the largest mound in the survey area.
Ceramic density around Structure 7 is 18 and is next to the densest
zone in the entire survey area, which, in Area 40a, has a density
figure of 673 sherds per hectare. Structure 9 is in Area 1, which was
tested internally by means of four-meter-sample circles. These samples
showed that ceramics were abundant at the west end of the area around
Feature 1 (a midden area), but as Structure 9 is approached at the
east end of the area, ceramics decrease markedly. Low density of
ceramics around the mounds may be attributable to two activities in
the past. If the mounds were a focus of higher-status activity, the
area around them may have been kept clean of rubbish. Another possibility is that construction of the mounds occurred shortly before
abandonment of the area, and fill used to build the structures was
scraped up from the surrounding area, removing most of the sherds
from the surface and incorporating them into the structures. This is
a distinct possibility for Structure 9, which, as discussed above,
probably was built toward the end of Late Monte Albán I (the area was
largely abandoned sometime during Monte Albán II). Also, when Structure 9 was built over Structure 9-sub-1, the surrounding terrace was
raised over a meter in height, which may have covered many sherds.
A similar explanation for low-density areas at Teotihuacán has been
suggested by Cowgill (1974).

Excavation also exposed several nonmound residences. Structure 14 was the only one in which a large enough area was exposed to enable its arrangement to be described. The floor of the structure was composed of hard clay which capped a layer of rough, angular stones. This floor surrounded a rectangular, unpaved patio, within which the remains of two individuals were found (Burials 2 and 3). Structure 14 is not the typical "open household cluster" found by Winter (1974b) in his excavations of Monte Albán I residences. Structure 14 more closely resembles the "closed household cluster" characteristic of the Classic period (ibid.). The other excavations were small, and only exposed parts of structures. Structure 13 consisted of a small area of rough stone flooring. Structures 11 and 12 are represented by partial walls, one of which (Structure 11) had an associated floor made of fitted flat limestone flagging. Structure 15 was a north-south wall of cut rectangular blocks which had a doorway, while Structures 16, 17, and 18 are superimposed constructions consisting of a large rock wall, an area of flagging similar to Structure 11, and a low platform of unknown dimensions composed of small limestone blocks. Feature 1 represents a midden area with three small pits (which may have been hearths) and was located near Structure 14, indicating a possible functional association with it. If so, the differences in frequencies of ceramic types between Feature 1 and Structure 14 discussed in chapter VII may be functionally significant, and certain fancier ceramic types may have been restricted to use in the house.

It is regrettable that excavation could not have been undertaken after the survey, rather than before. If such had been the case, other high-density areas could have been tested, especially in the Lower Sector. Also, the time limits imposed on the excavation rendered much of the excavation results inconclusive, since only small areas could be exposed.

It was mentioned above that Areas with densities greater than 100 rim sherds per hectare covered 6.5 hectares. In chapter XII it was inferred that Areas with Level 6 and above (densities greater than 100) were residential and that Level 5 areas (densities between 80 and 99) may also have been residential. The addition of Level 5 areas adds about 2.1 hectares, making a total of 8.6 hectares. Corrections for vegetation (excluding thornbush areas) indicate that vegetation may be masking another 3.3 hectares of area with densities greater than 100. If it is arbitrarily assumed that half of this area actually does have these higher densities, then another 1.6 hectares can be added to the occupational area, which brings the total residential area to 10.2 hectares during Late Monte Albán I.

It was estimated above that the Early Monte Albán I residential area was between 1.2 and 2.5 hectares, if Level 3 (density values between 30 and 50) areas were included along with vegetation corrections. This may seem inconsistent with Late Monte Albán I, where Level 5 or greater densities (density values above 80) are considered to be residential. This inconsistency is required if the Early Monte Albán I estimated residential area is to be a realistic one. Two factors may obviate this inconsistency. If the Early Monte Albán I occupation was of a shorter duration than the Late Monte Albán I occupation, a lower density figure for residential areas would be expected. It is possible that the Xoxocotlán piedmont was not colonized until well into the Early Monte Albán I phase. The second factor is that with much higher residential densities, as in Late Monte Albán I, the density of nonresidential debris will rise, since the absolute quantity of material being produced which can be spread around in nonhousehold areas will be much greater. Ceramic density was over eight times higher in Late Monte Albán I than in Early Monte Albán I, so that increasing the lower limit of residential density from 30 to 80 does not seem quite so inconsistent. If, on the other hand, a lower density figure for residential area during Late Monte Albán I is used, much of the ridge would be residential area, leaving little room for irrigation agriculture.

A density figure of number of persons per hectare is required to convert these measures of residential area into population estimates. Winter (1976:234) has suggested a figure of 10 households or about 50 people per two- or three-hectare Middle Formative village, giving a population density figure of about 20 persons per hectare. This is in the upper range of the density of Sanders' Low-Density Compact Villages with densities between 10 and 25 persons per hectare (Sanders 1965:50). An Early Monte Albán I residential area of between 1.2 and 2.5 hectares would have had an estimated population of between 24 and 50 people, assuming a density of 20 persons per hectare. If it is assumed that by the Late Formative, higher population densities obtained, then Winter's figure of 20 persons per hectare can be doubled, giving 40 persons per hectare, or a figure in the upper range of the densities Sanders suggests for High-Density Compact Villages (25 to 50 persons per hectare) (ibid.). Forty persons per hectare results in an Early Monte Albán I population of between 48 and 100 people. Applying the

206

same population density figures to the Late Monte Albán I residential area of between 6.5 and 10.2 hectares gives a population of between 130 and 204 at 20 persons per hectare, and between 260 and 408 at 40 persons per hectare.

The purpose of these population estimates is to compare them with the carrying capacity of the area and thereby determine whether a surplus would have been available for consumption by people living elsewhere on Monte Albán. The carrying capacity of the Xoxocotlán piedmont area, if irrigation is employed, has been discussed in chapter XIII, where it was concluded that the carrying capacity was somewhere between 1.25 and 5.0 persons per hectare of agricultural land. The lower figure takes into account the smaller corncob size in the past, but is based on rainfall agriculture, while the higher figure is based on the modern cob size using irrigation. It is probable that the Late Formative carrying capacity was closer to the higher figure than the lower, since the smaller cob size could have been compensated for by more intensive farming techniques such as terracing. Modern piedmont irrigators in the Valley do not employ terraces, whereas terraces are prevalent around the Late Formative Xoxocotlán canal system. Use of terraces concentrates water and soil nutrients in the fields instead of allowing them to rapidly drain off, and would have a significant effect on yields. Another possibility is communitywide coordination of crop cycles and fallow periods, which would decrease the amount of land required by any one family (see chapter XII).

The above discussion indicates that the carrying capacity was probably close to five persons per hectare, while the amount of land which could have been cultivated using water from the canal on the ridge was about the same as the area from which sherds were collected, or about 50 hectares. From this, the residential area (up to 10 hectares) must be subtracted, and land lost to erosion since the Late Formative must be added. Headward erosion of barrancas and arroyos has destroyed up to 10 hectares, so that 50 hectares is the best estimate for the amount of land cultivable during Late Monte Albán I, with slightly more land being available during Early Monte Albán I (due to the smaller residential area). Thus, the carrying capacity of the area was at most about 250 people (50 hectares times five persons per hectare).

The maximum population estimated for Early Monte Albán I was suggested to be approximately 100 people. This is well within the estimated carrying capacity, and indicates the possibility of producing a small surplus for consumption elsewhere on Monte Albán if the residents of the Xoxocotlán piedmont were motivated to do so and sufficient labor was available. However, during Late Monte Albán I, with an estimated population of between 130 and 400 people, the carrying capacity of the area was being approached, if not exceeded.

MONTE ALBAN II

The Xoxocotlán piedmont was largely abandoned during Monte Albán II. The abandonment of the area for settlement purposes may have been related to the pressure on the carrying capacity of the irrigation system or to other factors affecting Monte Albán as a whole. It appears that the Monte Albán II phase was a time of consolidation and compaction on Monte Albán. During Late Monte Albán I, areas off the main mountain were occupied: there was a small concentration of Late I sherds at the base of Monte Albán to the northwest (El Mogollito), and the tops of El Gallo and Monte Albán Chico were occupied, as was the Xoxocotlán piedmont. However, in Monte Albán II sherd density at El Mogollito and Xoxocotlán was substantially reduced, and El Gallo and Monte Albán Chico were abandoned (Blanton 1978:44). Also, what may have been a large defensive wall was constructed along the north and west sides of Monte Albán proper, possibly at the end of Late Monte Albán I or during Monte Albán II. This wall may also have served as an erosion-control device, retaining soil and runoff from the slopes above it for agriculture in large deep-soil terraces behind it. At one point, as the wall crosses an arroyo, it was strengthened to form a dam, thus creating a large reservoir (Neely 1972). Whatever its function, this wall was the boundary of Monte Albán II settlement. The abandonment of the Xoxocotlán irrigation settlement on the east side of Monte Albán may have been related to the abandonment of the outlying settlements to the north and northwest of Monte Albán proper and the construction of the wall in that area. The abandonment of peripheral areas and the construction of the wall may have been defensive measures of a weakened polity (Blanton 1978:55).

Although the Xoxocotlán piedmont was abandoned for residential purposes during Monte Albán II, it is possible that the canal continued to be used in later periods. Excavation revealed only Monte Albán II or earlier sherds in the canal fill (mostly Late Monte Albán I sherds), but there are so few sherds of later phases in the area that the canal fill could contain only Late Monte Albán I sherds even if the canal were in use until much later. The best evidence for the time of abandonment of the canal is the abandonment of the piedmont for residential purposes. It is not very likely that people who lived elsewhere would come to the area to cultivate irrigated crops. Irrigation is an intensive form of agriculture where proximity of residence to fields is the most economic solution to allocation of labor. This may, of course, have been offset by sociopolitical forces which kept the population nucleated on the upper slopes of Monte Albán.

LATER PERIODS

The Xoxocotlán Survey area was not occupied during the Classic period (Monte Albán III-A and III-B) except for a few small spots which may represent isolated houses. It is extremely difficult to distinguish between the ceramics from Monte Alban III-B and IV, but so few sherds from either Monte Albán III-B or IV were found in the Xoxocotlán piedmont that this problem is insignificant.

During Monte Albán V, the survey area was sparsely occupied, probably by people engaged in rainfall agriculture. Settlement elsewhere in the Valley at this time is dispersed, with one of the major settlement types being isolated houses. This seems to have been the pattern in the Xoxocotlán piedmont. Increases in corncob size probably were great enough by this time to make rainfall agriculture in the area practical if the population density remained low.

THE EFFECTS OF IRRIGATION

One of the reasons survey and excavation were undertaken in the Xoxocotlán piedmont area was to gather data on development of a Late Formative irrigation system and surrounding settlement. As stated in the Introduction, the two main objectives of the project were to determine effects of the irrigation system on the community pattern and to determine what contribution (if any) the irrigation system could have made to the food supply of Monte Albán, which is related to the scale of the system in terms of Wittfogel's (1957) hydraulic model, to be discussed below.

The community pattern has already been described above. In general, it appears that the effect of the irrigation system was to distribute residential areas so as not to interfere with agricultural activities. When expansion from the initial Early Monte Albán I settlement around Area 40a in the Lower Sector occurred, new settlement was confined to the Upper Sector, with its relatively rocky sloping soil. The Middle Sector, with its level surface and deeper soil, was apparently reserved for agriculture. This resulted in a nucleated residential area at the lower end of the canal and a relatively dispersed pattern of residence at the upper end of the canal, separated from each other by a largely agricultural area. It would appear that the pattern of original settlement in the Lower Sector was similar to that of Tierras Largas (Winter 1976:fig. 8.5) and Fabrica San José (Drennan 1976), during the Rosario phase where a roughly circular cluster of contiguous low-status residences contained a single high-status residence. The Xoxocotlán Lower Sector pattern more closely resembles the Fabrica San José pattern, where the high-status residence is at one edge of the community, as is Structure 7 at Xoxocotlán. However, when population increased and the Upper Sector was colonized, a more dispersed pattern of high- and low-status residences was formed.

What follows is a speculative reconstruction of the development of the community pattern discussed above. Ranking of lineages in the Valley of Oaxaca was probably well established by the end of the Middle Formative. This can be seen in differences in house construction and distribution of fancy objects at San José Mogoté and Huitzo (Flannery 1968, 1970, 1976b). Ranking probably applied to lineages and sublineages, not specifically to individuals. The colonization of the Xoxocotlán piedmont could have been the result of the budding-off of a lower-ranking lineage from a Valley-floor village which established itself in and around Area 40a in the Lower Sector during Early Monte Albán I. The irrigation system they constructed probably consisted of a small dam and a shallow ditch which brought water to parts of the Middle and Lower Sectors.

As population increased, budding-off occurred to the Upper Sector (which is steeper, poorer land) and more land was brought under cultivation. Probably sometime during Late Monte Albán I the dam was enlarged to its present size and terraces in the Upper Sector were constructed, possibly with the assistance of (or through the intervention of) the Monte Albán hierarchy.

The highest-ranking members of the original founding lineage may have been associated with Structure 7 in the Lower Sector. It is the largest mound in the area and is adjacent to the largest, densest, and probably earliest occupied residential area. As time went on, it was probably enlarged until it reached its present size at the end of Late

Monte Albán I. Mounds in the Upper Sector may represent residences of the heads of cadet or junior lineages which budded off from the Lower Sector. It is possible that disputes over water allocation could have developed between the Upper and Lower Sectors, which could have been settled by the intervention of the Monte Albán hierarchy.

The second objective of the Xoxocotlán Project was to determine what contribution the irrigation system could have made to the food supply of Monte Albán and what effect, if any, it had on the development of more complex political systems at Monte Albán in terms of the hydraulic hypothesis. As discussed above, during Early Monte Albán I, the population of the community around the canal was about 100, while the maximum carrying capacity of the system was 250 persons. Thus, a slight surplus (enough for 150 people) for the population of Monte Albán could have been produced. However, the population of Monte Albán during Early Monte Albán I is estimated by Blanton (1978:35) to have been between 3,500 and 7,000, making the contribution of the Xoxocotlán system insignificant. During Late Monte Albán I, the population of the Xoxocotlán community was probably close to the carrying capacity of the system (250 persons) and therefore probably could have contributed little, if anything, to the food supply of Monte Albán. Monte Albán at this time had a population of between 10,200 and 20,400, according to Blanton's estimate (Blanton 1978:44).

Thus, rather than "an attempt on the part of the community's [Monte Albán's] population to achieve more self-sufficiency" as Blanton (1978:55) suggests, the irrigation system probably represents a colonization of the piedmont below Monte Albán by Valley-floor agriculturalists who sought proximity to the new ceremonial-administrative center on Monte Albán. Construction of the irrigation system on the Xoxocotlán piedmont required a large labor investment, since the system had to be based on impoundment of runoff rather than diversion of a perennial stream. The dam had to be large and impermeable, rather than merely diversionary, and the distance between the dam and the best agricultural land was quite great, requiring a long canal (two kilometers). Thus, colonization of the Xoxocotlán piedmont by irrigation agriculturalists does not seem probable unless a stimulus in the form of a large ceremonial-administrative center were already present.

The hydraulic hypothesis suggests that the managerial requirements of irrigation agriculture, such as system construction and maintenance and water allocation, stimulate more complex political organization

(<u>cf</u>. Sanders 1968). However, in this case, where a small-scale system
was built requiring a relatively large labor investment, it appears
that the irrigation system was a response to the presence of an administrative center rather than vice versa.

The Xoxocotlán irrigation system is only one of a number of known
Formative period irrigation systems. The search for early irrigation
systems was stimulated by debate about the hydraulic hypothesis and
its applicability to the study of sociopolitical development in Mesoamerica (Armillas 1948; Palerm 1955; Millon 1962). Archaeological
research designed to locate irrigation systems was first carried out
in the Basin of Mexico by Millon, Armillas, and others (Millon 1954,
1957; Armillas, Palerm, and Wolf 1956; Wolf and Palerm 1955), but only
succeeded in locating systems of the Postclassic period near Teotihuacán
and Texcoco. In order to demonstrate causality or at least a systemic
relationship between irrigation and complex political organization,
it was necessary to locate irrigation systems which could be dated to
the Formative period, when the state and urbanism were developing.
Existence of water-control technology in the Formative period has now
been definitely established in the Tehuacán Valley (Woodbury and Neely
1972), the Puebla area (Fowler 1969), and in Oaxaca (Neely 1967, 1970,
1972). Canals have been noted at Cuicuilco (Palerm 1961), but at
Teotihuacán the evidence for Formative or Classic period irrigation is
still indirect (Price 1971). An irrigation system has recently been
found near Ecatepec in the Basin of Mexico which probably dates to
the Middle Formative period (Sanders and Santley 1977).

The Formative period irrigation system discovered in the Arroyo
Lencho Diego of the Tehuacán Valley was based on impoundment of
seasonal runoff, as was the Xoxocotlán system in Oaxaca, although
the Lencho Diego system had a much larger dam and reservoir. The
Lencho Diego system is also earlier than the Xoxocotlán system, dating
to 700 B.C. (Woodbury and Neely 1972). However, the Lencho Diego
system was not originally closely associated with a major settlement,
and it is not known why it was built in a seasonally flowing arroyo
(requiring a large dam for storage of summer runoff) when there is a
perennial stream in the next canyon to the south (the Tilapa). More
information on this system and its effect on nearby communities may be
forthcoming from the work of Spencer and Redmond in the area (Spencer
and Redmond 1977).

There is another irrigation system in Oaxaca, besides the
Xoxocotlán system, at Hierve el Agua, which is outside the Valley

of Oaxaca east of Mitla. This system had a travertine spring as its water source, from which a canal system led to a series of terraced fields (Neely 1967). This system is also not associated with any major settlement. The Xoxocotlán system, which is associated with the early stages of development of a major settlement (Monte Albán) has been shown to have been of such small scale that it probably had little effect on the development of Monte Albán. Thus, for the Southern Highlands of Mesoamerica, at any rate, there is little direct evidence supporting the hydraulic hypothesis. Better evidence for the relationship of irrigation to the development of complex political institutions may come from Amalucan in Puebla (Fowler 1969) and Cuicuilco in the Basin of Mexico (Palerm 1961), where irrigation systems are associated with larger communities. However, as yet there has been no published study on the relative sizes of these systems and their associated settlements, as has been done for the Xoxocotlán system. At Cuicuilco, this may not be possible because of the lava covering most of the site.

Lees (1973:94) argues from settlement-pattern evidence that in the Valley of Oaxaca during the Late Formative period, perennial streams flowing into the Valley were used as water sources for diversionary canal irrigation systems. Lees notes that during the Late Formative, what seem to be hilltop administrative centers appear overlooking these perennial streams in the upper piedmont, where sites had previously been rare. Lees suggests that these new hilltop sites may have functioned to allocate water and settle disputes in the hypothesized irrigation systems. A similar case could be made for Tilapa Canyon in the Tehuacán Valley, where hilltop centers are found on either side of the perennial stream at the upper end of the canyon dating to the beginning of the Classic period, when the nearby Lencho Diego system was abandoned (Woodbury and Neely 1972, McNeish et al. 1975, and reconnaissance by Mason and O'Brien in 1974).

However, inferring an irrigation system from the location of hill-top settlements and then inferring that the irrigation system caused the hilltop settlements is circular reasoning. Before the hydraulic hypothesis can be tested further in this region, the irrigation systems associated with the hilltop centers must be found, and the relative scales of the irrigation system and associated settlements must be determined, as was done for the Xoxocotlán system.

Possible effects of irrigation include: increasing the yield of a given amount of land, which allows colonization of previously un-

cultivable areas; encouraging settlement nucleation around the irrigation system; restructuring the internal community pattern to conform
to the structure of the irrigation system; and promoting development
of more complex forms of political organization (the hydraulic
hypothesis). The Xoxocotlán irrigation system provides evidence for
the first three effects listed. As discussed above, the Xoxocotlán
piedmont could not have been cultivated without irrigation, due to the
small corncob size during the Late Formative period (Kirkby 1973).
Irrigation allowed colonization of the area, making agriculture possible
near the Monte Albán regional center. The irrigation system did
nucleate settlement around the canal, in that no areas off the ridge
containing the irrigation system were occupied. However, the internal
community pattern was relatively dispersed, with a large residential
area at the lower end and several smaller residential areas at the
upper end of the canal. The effect of the irrigation system on the
internal community pattern was to promote a residential pattern
parallel to the canal on land less suitable for agriculture.

The irrigation system, because of its small size, could have had
little effect on the political development of Monte Albán (the fourth
possible effect listed above). However, it is possible that the
irrigation system did have an effect on the internal political organization of the community associated with the canal. The only evidence
for this is what seems to be a large number of mounds for the residential area present. This may indicate a larger-than-average residential
elite group. However, until other nonirrigation Late Formative
communities are studied to determine what the "average" ration of
mounds to residential area is, the above remains speculative.

References Cited

Armillas, Pedro
    1948 A sequence of cultural development in Mesoamerica. In A
           reappraisal of Peruvian archaeology, edited by Wendall C.
           Bennett. Memoirs of the Society for American Archaeology
           4.

Armillas, Pedro, Angel Palerm, and Eric R. Wolf
    1956 A small irrigation system in the Valley of Teotihuacán.
           American Antiquity 21:296-299.

Bass, William M.
    1971 Human osteology: A laboratory and field manual of the human
           skeleton. Missouri Archeological Society, Columbia, Missouri.

Bernal, Ignacio
    1947 La cerámica preclásica de Monte Albán. Thesis, Escuela
           Nacional de Antropología e Historia, Mexico City.
    1949a La cerámica grabada de Monte Albán. Anales del Instituto
           Nacional de Antropología e Historia 3:59-78.
    1949b La cerámica de Monte Albán III-A. Thesis, Universidad Nacional
           Autónoma de México.
    1965 Archaeological synthesis of Oaxaca. Handbook of Middle American
           Indians 3:788-813. University of Texas Press, Austin.

Blanton, Richard E.
    1972 Prehispanic settlement patterns of the Ixtapalapa Peninsula
           region, Mexico. Occasional Papers in Anthropology 6.
           Department of Anthropology, The Pennsylvania State University,
           University Park.
    1973 The Valley of Oaxaca settlement pattern project. Report
           submitted to the National Science Foundation and the Instituto
           Nacional de Antropología e Historia, Mexico.
    1978 Monte Albán: settlement patterns at the ancient Zapotec
           capital. Academic Press, New York.

Brennan, Curtiss
    1971 Modern Oaxacan agriculture and Prehispanic Mesoamerican
           subsistence systems. Manuscript on file, Department of
           Anthropology, University of Texas.

Brumfiel, Elizabeth
    1976 Specialization and exchange at the Late Postclassic (Aztec)
           community of Huexotla, Mexico. Ph.D. dissertation, University
           of Michigan. University Microfilms, Ann Arbor.

216

Byland, Bruce E.
    1978   Boundary recognition in the Mixteca Alta, Oaxaca, Mexico.
           Paper presented at the 43rd annual meeting of the Society
           for American Archaeology, Tucson.

Caso, Alfonso
    1938   Exploraciones en Oaxaca, quinta y sexta temporadas, 1936-37.
           Publicaciones del Instituto Panamericano de Geografía e
           Historia 34.

Caso, Alfonso and Ignacio Bernal
    1952   Urnas de Oaxaca. Memorias del Instituto Nacional de Antro-
           pología e Historia 2.

Caso, Alfonso, Ignacio Bernal, and Jorge R. Acosta
    1967   La cerámica de Monte Albán. Memorias del Instituto Nacional
           de Antropología e Historia 13.

Cowgill, George
    1974   Quantitative studies of urbanization at Teotihuacán. In
           Mesoamerican archaeology: new approaches, edited by Norman
           Hammond, pp. 363-396. University of Texas Press, Austin,
           Texas.

Dancey, William S.
    1973   Prehistoric land use and settlement pattern in the Priest
           Rapids area, Washington. Ph.D. dissertation, University of
           Washington. University Microfilms, Ann Arbor.

Drennan, Robert D.
    1976   Fábrica San José and Middle Formative society in the Valley
           of Oaxaca, Mexico. Memoirs of the Museum of Anthropology 8.
           University of Michigan.

Dunnell, Robert E. and William S. Dancey
    n.d.   Siteless surveys: a regional data collection strategy.
           In The design of archaeological research, edited by L. Mark
           Raab and Timothy C. Klinger. Aldine Press, Chicago. In
           press.

Flannery, Kent V.
    1968   The Olmec and the Valley of Oaxaca: a model for interregional
           interaction in Formative times. In Dumbarton Oaks conference
           on the Olmec, edited by Elizabeth P. Benson, pp. 79-110.
    1976a Research strategy and Formative Mesoamerica. In The early
           Mesoamerican village, edited by Kent V. Flannery, pp. 1-12.
           Academic Press, New York.
    1976b Sampling by intensive surface collection. In The early
           Mesoamerican village, edited by Kent V. Flannery, pp. 51-62.
           Academic Press, New York.

Flannery, Kent V. (editor)
    1970   Preliminary archaeological investigations in the Valley of
           Oaxaca, Mexico, 1966-69. Report to the National Science
           Foundation and the Instituto Nacional de Antropología e
           Historia, Mexico.
    1976   The early Mesoamerican village. Academic Press, New York.

Flannery, Kent V., Anne Kirkby, Michael Kirkby, and Aubrey Williams
1967   Farming systems and political growth in ancient Oaxaca. Science 158:445-454.

Flannery, Kent V. and James Schoenwetter
1970   Climate and man in Formative Oaxaca.   Archaeology 23:144-152.

Ford, James A.
1962   A quantitative method for deriving cultural chronology. Technical Manual 1.   Pan American Union, Washington, D.C.

Fowler, Melvin
1969   A preclassic water distribution system in Amalucan, Mexico. Archaeology 22:208-215.

Granskog, Jane
1974   Efficiency in a Zapotec Indian agricultural village.   Ph.D. dissertation, University of Texas at Austin.   University Microfilms, Ann Arbor.

Hirth, Kenneth G.
1974   Pre-columbian population development along the Rio Amatzinac: the Formative through Classic periods in eastern Morelos, Mexico.   Ph.D. dissertation, University of Wisconsin. University Microfilms, Ann Arbor.
1976   Surface reconnaissance in the Coatlán-Tetecala region, Morelos, Mexico.   Report submitted to the Instituto Nacional de Antropología e Historia, Mexico.

Hole, Frank, Kent V. Flannery, and James A. Neely
1969   Prehistory and human ecology of the Deh Luran Plain.   Memoirs of the Museum of Anthropology 1.   University of Michigan, Ann Arbor.

Kirkby, Anne V. T.
1973   The use of land and water resources in the past and present Valley of Oaxaca, Mexico.   Memoirs of the Museum of Anthropology 5.   University of Michigan, Ann Arbor.

Kolb, Charles C.
1973   Thin orange pottery at Teotihuacán.   Occasional Papers in Anthropology 8:309-377.   Department of Anthropology, The Pennsylvania State University, University Park.

Kowalewski, Stephen
1976   Prehispanic settlement patterns of the central part of the Valley of Oaxaca.   Ph.D. dissertation, University of Arizona. University Microfilms, Ann Arbor.
1978   Growth and non-growth in the past and present Valley of Oaxaca, Mexico.   Paper presented at the 43rd annual meeting of the Society for American Archaeology, Tucson.

Kowalewski, Stephen A., Charles Spencer, and Elsa Redmond
1978   Description of ceramic categories.   Appendix II of Monte Albán: settlement patterns at the ancient Zapotec capital, by Richard E. Blanton, pp. 167-193.   Academic Press, New York.

218

Kuttruff, Carl
1978 Figurines and urn fragments from the Monte Albán survey.
Appendix VIII of Monte Albán: settlement patterns at the
ancient Zapotec capital, by Richard E. Blanton, pp. 379-402.
Academic Press, New York.

Lees, Susan H.
1973 Socio-political aspects of canal irrigation in the Valley of
Oaxaca, Mexico. Memoirs of the Museum of Anthropology 6.
University of Michigan, Ann Arbor.

Lewarch, Dennis E. and Roger D. Mason
1977 Proyecto Coatlán: Surface collection and excavation in the
Coatlán del Río Valley, Morelos, Mexico. Report submitted to
Instituto Nacional de Antropología e Historia, Mexico.

MacNeish, Richard S., Melvin L. Fowler, Angel Garcia Cook, Frederick
A. Peterson, Antoinette Nelken Turner, and James A. Neely.
1975 The prehistory of the Tehuacán Valley, vol. 5: Excavation
and reconnaissance. University of Texas Press, Austin.

MacNeish, Richard S., Frederick A. Peterson, and Kent V. Flannery
1970 The prehistory of the Tehuacán Valley, Vol. 3: Ceramics.
University of Texas Press, Austin.

Mason, Roger D.
1979 Regional, zonal, and site-intensive surveys in central
Mexico. Western Canadian Journal of Anthropology 8:89-105.

Mason, Roger D., Dennis E. Lewarch, Michael J. O'Brien, and James A.
Neely
1977 An archaeological survey on the Xoxocotlán piedmont, Oaxaca,
Mexico. American Antiquity 42:567-575.

Millon, René
1954 Irrigation at Teotihuacán. American Antiquity 20:177-180.

1957 Irrigation systems in the Valley of Teotihuacán. American
Antiquity 23:160-166.
1962 Variations in the social responses to the practice of irrigation agriculture. In Civilizations in arid lands, edited
by Richard Woodbury, pp. 56-88. University of Utah Anthro-
pological Papers 62.

Neely, James A.
1967 Organización hydráulica y sistemas de irrigación prehistóricos
en el Valle de Oaxaca. Boletín 27:15-17. Instituto Nacional
de Antropología e Historia, Mexico City.
1970 Terrace and water control systems in the Valley of Oaxaca
region. In Preliminary archaeological investigations in the
Valley of Oaxaca, Mexico, edited by Kent V. Flannery. Report
submitted to the National Science Foundation and the Instituto
Nacional de Antropología e Historia, Mexico.
1972 Prehistoric domestic water supplies and irrigation systems at
Monte Albán, Oaxaca, Mexico. Paper presented at the 38th
annual meeting of the Society for American Archaeology, Miami.

Neely, James A. and Michael J. O'Brien
    1973 Irrigation and settlement nucleations at Monte Albán: a test
          of models. Paper presented at the 39th annual meeting of the
          Society for American Archaeology, San Francisco.

Nie, Norman, Dale H. Bent, and C. Hadlai Hull
    1970 SPSS: Statistical package for the social sciences. McGraw-
          Hill, New York.

Orlandini, Richard J.
    1967 A Formative well from the Valley of Oaxaca. Paper presented
          at the 32nd annual meeting of the Society for American Archae-
          ology, Boston.

Paddock, John
    1962 Notes on the Caso and Bernal typology of Monte Albán pottery.
          Manuscript, Mexico City College, Mexico.

Palerm, Angel
    1955 The agricultural bases of urban civilization in Mesoamerica.
          In Irrigation civilizations: a comparative study, edited by
          Julian H. Steward, pp. 28-42. Social Science Monographs 1.
          Pan American Union, Washington, D.C.
    1961 Sistemas de regadío prehispánico en Teotihuacán y en el
          Pedregal de San Angel. Revista Interamericana de Ciencias
          Sociales, época 2, vol. 1:297-302.

Parsons, Jeffrey R.
    1971a Prehistoric settlement patterns in the Texcoco region, Mexico.
          Memoirs of the Museum of Anthropology 3. University of
          Michigan, Ann Arbor.
    1971b Prehispanic settlement patterns of the Chalco region, Mexico.
          Report submitted to the Instituto Nacional de Antropología
          e Historia, Mexico.
    1974 The development of a prehistoric complex society: a regional
          perspective from the Valley of Mexico. Journal of Field
          Archaeology 1:81-108.

Pires-Fereira, Jane
    1975 Formative Mesoamerican exchange networks, with special refer-
          ence to the Valley of Oaxaca. Memoirs of the Museum of Anthro-
          pology 7. University of Michigan, Ann Arbor.

Price, Barbara
    1971 Prehispanic irrigation agriculture in nuclear America.
          Latin America Research Review 6:3-60.

Sanders, William T.
    1965 The cultural ecology of the Teotihuacán Valley. Department
          of Anthropology, The Pennsylvania State University, University
          Park.
    1968 Hydraulic agriculture, economic symbiosis, and the evolution
          of states in central Mexico. In Anthropological archaeology
          in the Americas, edited by Betty J. Meggers, pp. 88-107.

220

1970   The population of the Teotihuacán Valley, Basin of Mexico,
       and the Central Mexican Symbiotic Region in the sixteenth
       century. In The Teotihuacán Valley Project, Final Report,
       vol. 1, Natural environment, contemporary occupation, and
       16th century population of the Valley, edited by William T.
       Sanders et al., pp. 385-457. Occasional Papers in Anthropology
       3. Department of Anthropology, The Pennsylvania State Uni-
       versity, University Park.

Sanders, William T. et al. (editors)
1970   The Teotihuacan Valley Project, Final Report, vol. 1,
       Natural environment, contemporary occupation, and 16th century
       population of the Valley. Occasional Papers in Anthropology
       3. Department of Anthropology, The Pennsylvania State University,
       University Park.
1975   The Teotihuacán Valley Project, Final Report, vol. 2, The
       Formative period. Occasional Papers in Anthropology 10.
       Department of Anthropology, The Pennsylvania State University,
       University Park.

Sanders, William T. and Robert S. Santley
1977   A Prehispanic irrigation system near Santa Clara Xalostoc
       in the Basin of Mexico. American Antiquity 42:582-588.

Saville, Marshall
1899   Exploration of Zapotecan tombs in southern Mexico. American
       Anthropologist 1:350-362.

Shepard, Anna O.
1967   Preliminary notes on the paste composition of Monte Albán
       pottery. Appendix to La cerámica de Monte Albán by Alfonso
       Caso et al. Memorias del Instituto Nacional de Antropología
       e Historia 13.

Schmidt, Allan
1973   Synagraphic mapping program (SYMAP). Lab-Log 3-5. Laboratory
       for Computer Graphics and Spatial Analysis, The Graduate School
       of Design, Harvard University, Cambridge.

Sisson, Edward B.
1973   First annual report of the Coxcatlán Project. Robert S. Pea-
       body Foundation for Archaeology, Andover, Massachusetts.

Spencer, Charles S. and Elsa M. Redmond
1977   Survey in the arroyo Lencho Diego. In The Palo Blanco Project,
       a report on the 1975 and 1976 seasons in the Tehuacán Valley,
       edited by Robert D. Drennan. R. S. Peabody Foundation for
       Archaeology, Andover, Massachusetts.

Spores, Ronald
1972   An archaeological settlement survey of the Nochixtlán Valley,
       Oaxaca. Vanderbilt University Publications in Anthropology 1.

Tolstoy, Paul R. and Suzanne K. Fish
1975   Surface and subsurface evidence for community size at
       Coapexco, Mexico. Journal of Field Archaeology 2:97-104.

Trotter, Mildred and Goldine Glesser
  1952 Estimation of stature from long bones of American Whites and
        Negroes. American Journal of Physical Anthropology 10:463-514.
  1958 A re-evaluation of estimation of stature based on measurements
        of stature taken during life and of long bones after death.
        American Journal of Physical Anthropology 16:79-123.

Varner, Dudley M.
  1974 Prehispanic settlement patterns in the Valley of Oaxaca,
        Mexico: the Etla arm. Ph.D. dissertation, University of
        Arizona. University Microfilms, Ann Arbor.

Winter, Marcus C.
  1970 Excavations at Tierras Largas (Atzompa, Oaxaca): a preliminary
        report. In Preliminary archeological investigations in the
        Valley of Oaxaca, Mexico, edited by Kent V. Flannery. Report
        submitted to the National Science Foundation and the Instituto
        Nacional de Antropología e Historia, Mexico.
  1972 Tierras Largas: A Formative community in the Valley of
        Oaxaca, Mexico. Ph.D. dissertation, University of Arizona.
        University Microfilms, Ann Arbor.
  1974a Late Formative society in the Valley of Oaxaca and the Mixteca
        Alta, Mexico. Paper presented at the 41st meeting of the
        International Congress of Americanists, Mexico City.
  1974b Residential patterns at Monte Albán, Oaxaca, Mexico. Science
        186:981-987.
  1976 Differential patterns of community growth in Oaxaca. In
        The early Mesoamerican village, edited by Kent V. Flannery,
        pp. 117-234. Academic Press, New York.

Winter, Marcus C., Margarita Gaxiola, and Adriana Alaníz
  1975 Secuencia arqueológica del Valle de Oaxaca. Sección de
        arqueología, Centro Regional de Oaxaca, Instituto Nacional
        de Antropología e Historia, Mexico.

Wittfogel, Karl
  1957 Oriental despotism: a comparative study of total power.
        Yale University Press, New Haven.

Wolf, Eric R. and Angel Palerm
  1955 Irrigation in the old Acolhua domain, Mexico. Southwestern
        Journal of Anthropology 11:265-281.

Woodbury, Richard B. and James A. Neely
  1972 Water control systems of the Tehuacán Valley. In Prehistory
        of the Tehuacán Valley, vol. 4, chronology and irrigation,
        edited by Frederick Johnson, pp. 81-153. University of
        Texas Press, Austin.

PLATES

Plate 1. View of the Xoxoctlán Piedmont (looking south)

Plate 2. Main dam in the large barranca to the east of the South Platform, Monte Albán.

226

Plate 3. Trench through terrace wall surrounding Structure 9. In the background, trench continues through Structure 9 (looking south)

Plate 4. Structure 11, north-south wall and flagging

Plate 5. Profile through Structure 9, illustrating plaster floor (looking west)

Plate 6. Closeup of plaster floor, Structure 9; hearth in southwest corner

Plate 7. Plaster exterior of Structure 9 with Burial 1 on second tier (note remains of tomb on right side of body; looking north)

Plate 8. Burial 1

Plate 9. Area 1, Feature 1 (note pits in foreground)

Plate 10. Structure 14 with patio and Burials 2 and 3 in foreground (looking west)

Plate 11. Large retaining wall facing the east slope of Area 7

Plate 12. Structure 15 (looking north)

Plate 13. Structure 16 (looking north).

Plate 14. Structure 17, flagging

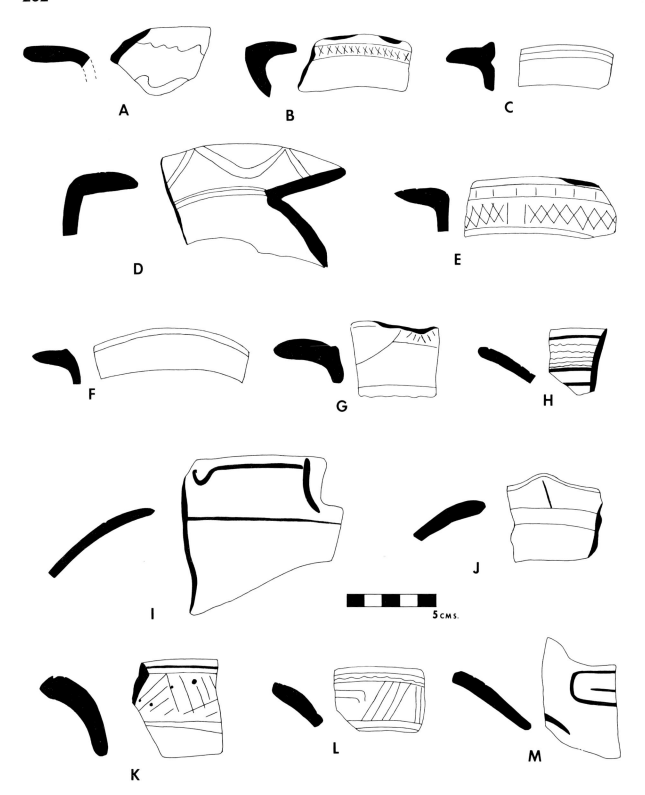

Plate 15. Early Monte Albán rim forms. A-G, 90° rims; H-L, everted rims; M, eccentric rim

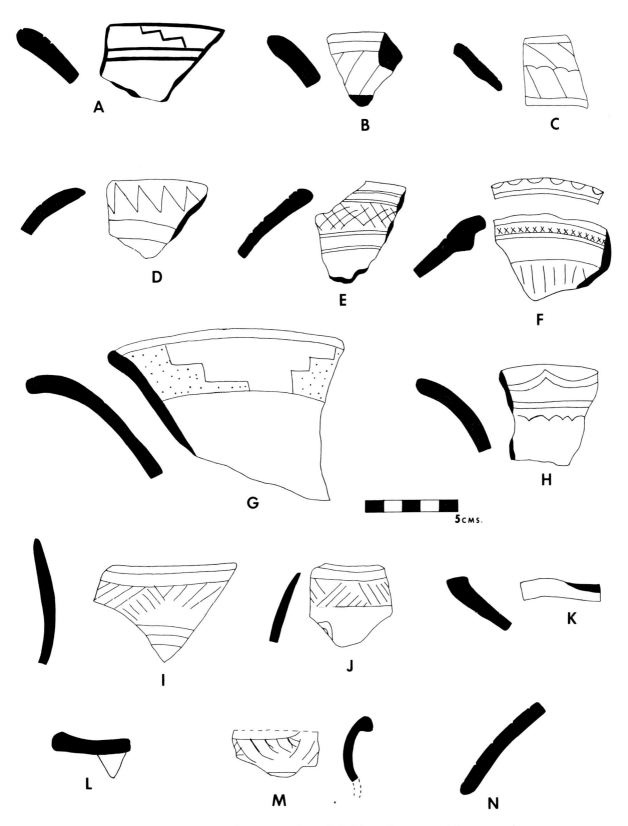

Plate 16. Early Monte Albán I ceramics. A-J, M, various vessel forms and accompanying designs; K, pinched-rim cajete; L, comal; N, pot stand.

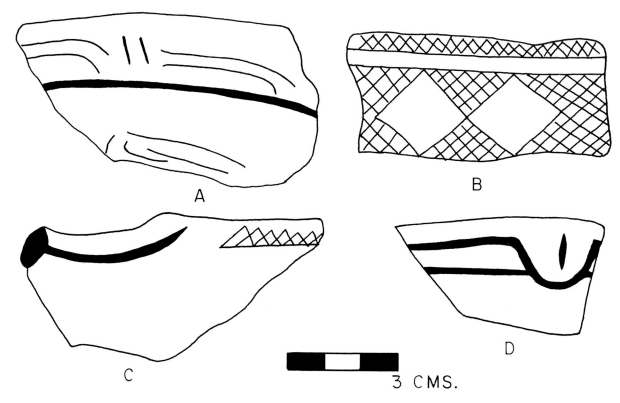

Plate 17. Early Monte Albán I design elements. A, B, exterior incising; C, D, eccentric-rimmed vessels

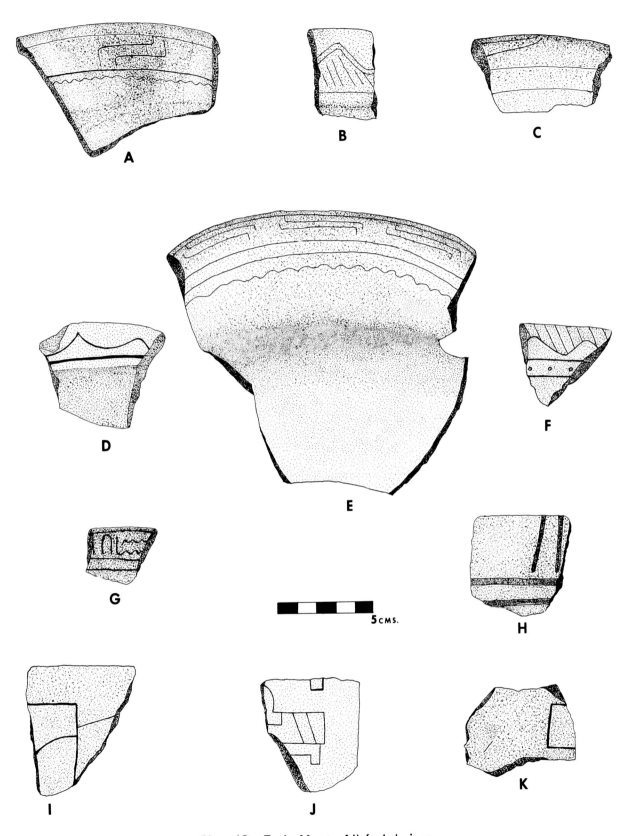

Plate 18. Early Monte Albán I designs

236

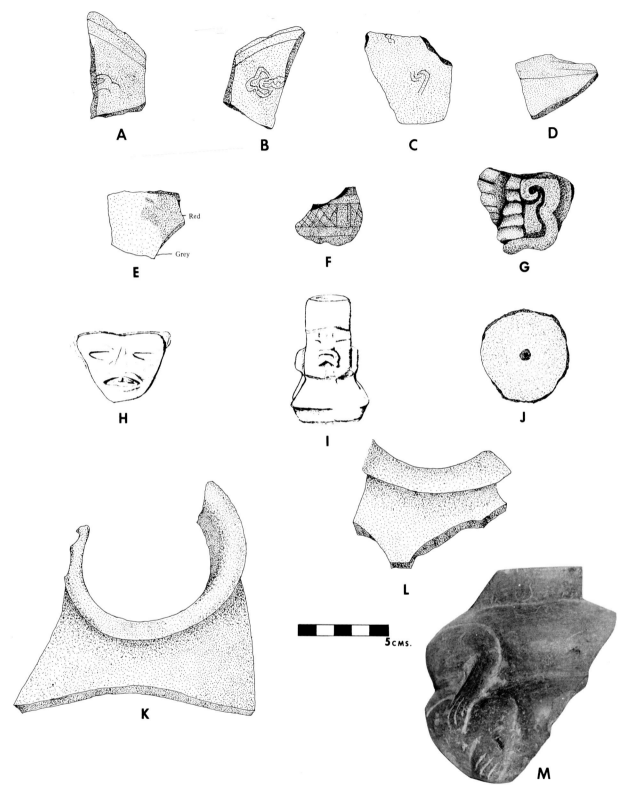

Plate 19. Early and Late Monte Albán I ceramics. A-D, F, external fine-line
incising; E, red wash; G, urn fragment (Monte Albán III-B); H, I,
figurines; J, pottery disc; K-L, fine gray ware

237

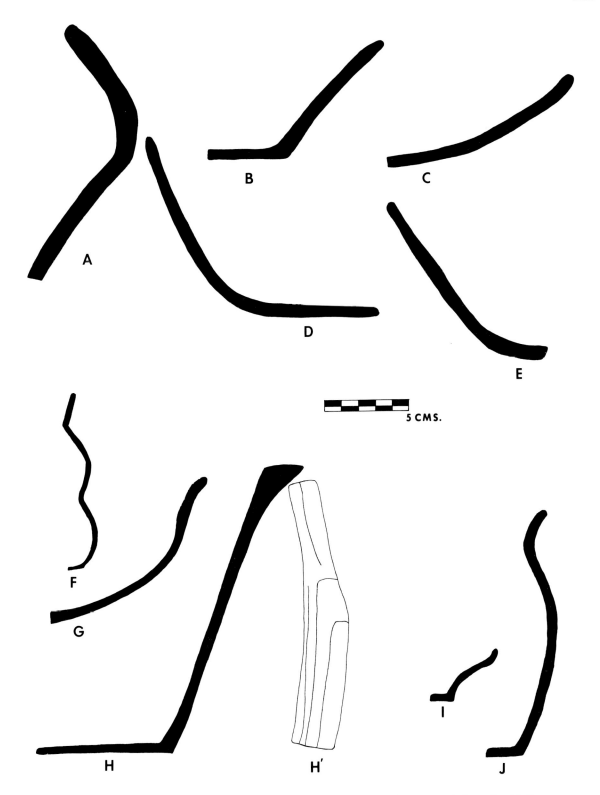

Plate 20. Early and Late Monte Albán I vessel forms. A, overted rim olla; B-E, G, I, various style cajetes; F, silhouette of vessel shown in pl. 19 (M); H, H´, silhouette and rim design of Early Monte Albán I illustrated in pl. 21 (A); J, small olla

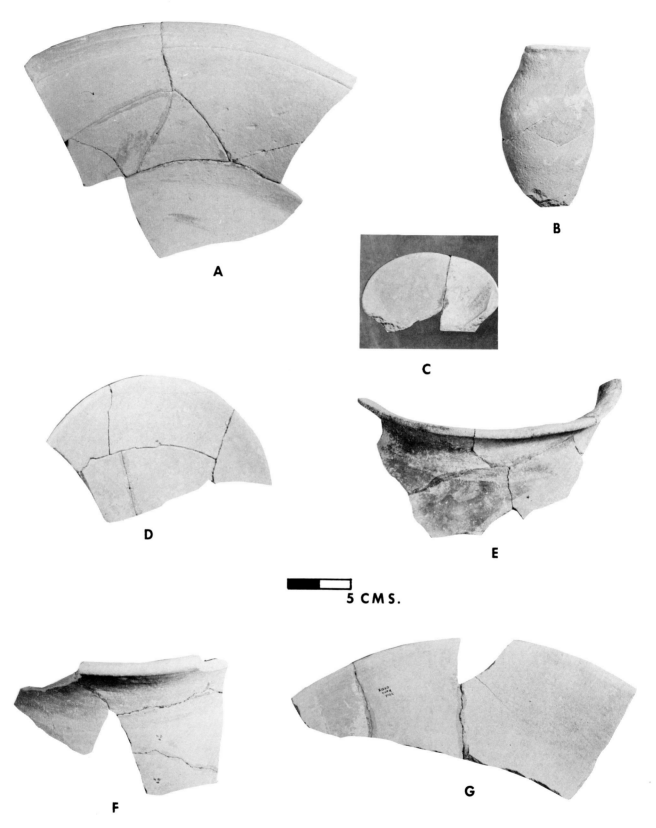

Plate 21. Various vessel forms. A, Early Monte Albán I 90°-rim basin; B, Monte
Albán II? vase; C, Late Monte Albán I everted-rim cajete; D, G, Late
Monte Albán I plates or comales; E, F, Late Monte Albán I ollas

239

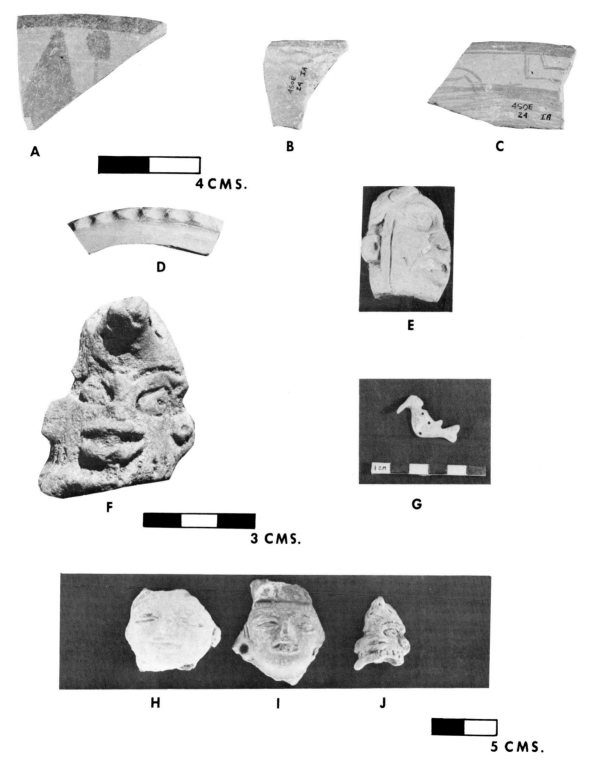

Plate 22. Early Monte Albán I rims, various style figurines, and drilled shell pendant.
A-C, Early Monte Albán I burnished rims; D, Early Monte Albán I scalloped
rim; E, F, H-J, various style figurines; G, drilled shell pendant

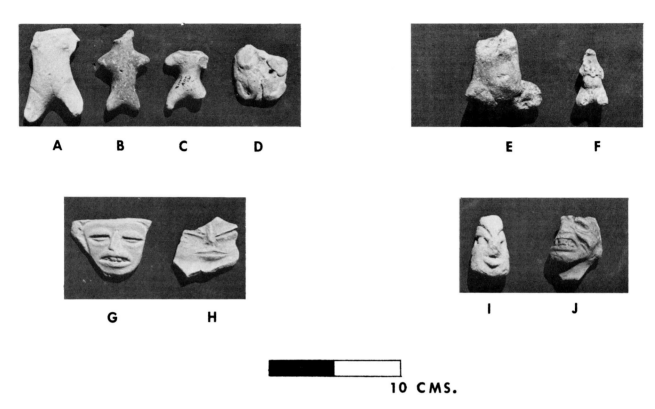

A   B   C   D   E   F

G   H   I   J

10 CMS.

Plate 23. Various style figurines